A HUNDRED YEARS OF GEOGRAPHY

T. W. FREEMAN

A HUNDRED YEARS OF GEOGRAPHY

Routledge
Taylor & Francis Group

LONDON AND NEW YORK

First published 1961 by Transaction Publishers

Published 2017 by Routledge
2 Park Square, Milton Park, Abingdon, Oxon OX14 4RN
711 Third Avenue, New York, NY 10017, USA

Routledge is an imprint of the Taylor & Francis Group, an informa business

Library of Congress Catalog Number: 2006052040

Library of Congress Cataloging-in-Publication Data

Freeman, Thomas Walter.
   A hundred years of geography / T. W. Freeman
        p. cm.
   Includes bibliographical references and index.
   ISBN 978-0-202-30920-0 (alk. paper)
      1. Geography—History. I. Title.

G99.F7   2007
910.9'034—dc22

2006052040

ISBN 13: 978-0-202-30920-0 (pbk)

For
Gerald R. Crone

CONTENTS

PREFACE

THE history of geography is not an over-tilled field, and this book deals only with some aspects of its development in the past hundred years. The request to write it came by post with no preliminary warning but never was a task more happily accepted, for it offered an opportunity of reading much that was written in the nineteenth century and of observing the steady growth of geographical work in a rapidly changing world. One wonders, for example, how great the influence of geographical arguments on the Treaty of Versailles really was, and to what extent geopolitical thinking paved the way for the war of 1939–45. The impetus given to geography by colonial expansion in the late nineteenth century is clear, and it cannot be accidental that the growth of geography has been notable since 1919. Much could be done by a series of national geographical histories, of which one for the United States is promised already. Equally, there is a need for more biographies of geographers of various periods.

No subject, perhaps, can more justly claim to be international, and though this book is written with a basis of British geography, the debt of British geographers to the continental European pioneers, particularly in France and Germany, is heavy. In recent years, the marked advance of American geography has been fruitful in Europe: among the smaller nations fine work has been done, conspicuously by the Finns and Swedes. For the future, many hopes rest with the university and other geographers in Asia, Africa, South America, Australia and New Zealand, not least because their home territories are being rapidly changed and undoubtedly will be transformed in the next few generations.

It is usual for an author to thank his colleagues and other friends for help, but the main acknowledgement is to those who have facilitated escape into the monastic seclusion of library, study and garden. It is only fair to add that the views expressed are personal, and do not implicate the friend to whom this book is affectionately dedicated. A word of gratitude must be given to the small Honours class in Manchester who patiently listened to this book as a lecture course; possibly this book will help students of the day and graduates of an earlier time to see that there is no binding orthodoxy of view, no one geographer who was always right but rather a number of faithful workers and thinkers. If this book had a sub-title, it might be 'no new idea under the sun' for many ideas are produced, ignored, and revived fifty years or so later with good results. But such a sub-title would be unfair—for new methods and ideas are constantly being tried. And there is the hope for the future.

T. W. FREEMAN.

The University, Manchester.
January 1961

CHAPTER ONE

CHANGING GEOGRAPHY

A century of progress; six trends of geography; specialization and generalization

ANYONE who earns his living by teaching geography has to endure the comment that his subject is 'new', though in fact it goes back to the beginnings of learning as many historians of geography have shown. Its roots lie in the natural curiosity of people about places and ways of living other than their own, and at least from the days of Herodotus explorers and military conquerors wrote down what they saw for the benefit of governments and of a wider circle of readers. Speculation about the nature of the world, its shape, size and qualities goes back to the ancient Egyptians who viewed the sky as a kind of ceiling supported above the earth by four pillars corresponding to the cardinal points. In the third century B.C., Eratosthenes of Alexandria accepted the Greek view that the earth was a sphere with a diameter of some 25,000 miles.

The woes of Galileo and the fears of Columbus's sailors came from the medieval belief in a flat earth, but the darkness of the Dark Ages is often exaggerated for in fact the knowledge of the world was increasing all the time: explorers seeking conquest, trade or merely adventure, went forth and left accounts of their journeys and observations for posterity. It is not with the thousands of years of geography that this book deals—indeed, there are already fine histories available[1]—but rather with the last hundred years only. At the outset, however, it is well to realize that so much has gone before: during the past hundred years, more and more people have referred to themselves as 'geographers', but in many past generations people have been geographers in fact if not in name. One need pay little attention to those who designate themselves as belonging to the third (or second) generation of British geographers, for the shades of Hakluyt, Mary Somerville and many more stands in dumb rebuke.

A Century of Progress

The past hundred years have seen a vast growth of geographical knowledge. This has come through the opening-up of the world by conquest, trade, missionary enterprise and exploration, and above all through the provision of quick transport by steamship, railway and aeroplane. Within a century the population of the world has been doubled, vast new lands have been settled, the political maps altered almost beyond recognition, and new ideologies given practical expression in government and allied social policies. One may question the validity of the view of a British geographer, C. B. Fawcett, that the last hundred years are of more significance than all previous history, yet one is bound to recognize that the changes have been revolutionary. In the Rede Lecture of 1958,[2] Sir Charles Darwin noted that more minerals had been removed from the earth during the past forty years than in all previous time and that though the world's farmlands were producing more, the increase in food had not kept pace with the increase in people. It may be a difficult world to live in, but it is hardly a dull one.

Against such a background a great mass of raw material has been provided for geographical study. Raw material is always raw, and its discriminating use has depended on the growth of education in schools and universities and on the provision of scholars to use the rich resources available. The talented amateur, the critically-minded explorer, the natural scholar of independent means, have all existed and become known as geographers, but the real modern growth of the subject came with the recognition given by universities, in not a few cases reluctantly and even under the pressure of fashion—even universities keep up with the Joneses. Many of the world's early geographical societies regarded educational advance as essential to their work, though their main purpose was to encourage exploration and to gather up the fruits of enterprising penetration of the remoter areas of the world. Some, however, had other aims—in fact, as shown in chapter three there is no such thing as a standard geographical society. Perhaps it is just as well. Geography is by no means unique in its recent penetration of many of the world's universities; many other subjects, notably economics and the social studies, have only a comparatively short history of university recognition, and as shown on page 19, there have been demands for greater facilities for study by those responsible for several subjects in Britain quite recently. New opportunities were offered for geographical work, at varying times in different countries, as part of a general broadening of university education.

In the pages that follow, an attempt has been made to give an outline history of the modern growth of geography with a discussion of various aspects of the subject. It is not proposed to solve all the great controversies that have arisen, nor to advocate any particular view or doctrine: many of these are of considerable interest, even fascination, and some of them are obviously rooted in a divergence of views on life in general. In our own times both Fascism and Communism have been geographically expressed: the former brought arguments for spread of a greater Germany and the latter is prominent in Russian economic geographies which maintain how vastly beneficial all the rearrangements of population and the undeniable intensification of economic activity must be. Looking further back, the differences of outlook between Ritter and von Humboldt have been ascribed partly to the former's conception of a divine purpose in all existence, and the latter's more cautious and in some ways neutral approach to theological problems. Much of the stimulus to geographical inquiry in the nineteenth century came from the Darwinian hypothesis, and especially from the idea of the adaptation of organisms to environment with varying success: inspiration was given, too, by the widening of scientific enterprise, particularly in field study. Mackinder[3] has claimed that one of the main foundations of Darwin's work was the appreciation of the geographical distribution of animals under varied climatic conditions. Some organisms have successfully adapted themselves to changed circumstances, while others have not, and in time this had human analogies; for some stocks have apparently shown greater powers of adaptation in new areas, or under changed climates, than others. It is a truism that human life and environment have been intimately interwoven, biologically and culturally, from the beginning of life on earth, but extreme claims have been made for environmental influence, notably by Ellsworth Huntington in his studies of the effects of climate on human communities, or by some of the more vigorous writers on the effects of particular types of physical setting, such as mountains, plains, peninsulas or islands or social and political organization.

Such arguments are tempting. An American writer has shown that Finnish immigrants have been successful and happy in areas similar to those of their homeland, with coniferous forests, number-less lakes and rivers, and a cold winter.[4] Chisholm[5] in 1916 drew attention to the dangers of generalizations: he quoted a statement that 'nations that are accustomed to a limited territory, as were the Greeks, always search for a similar limited area' by pointing out that the Greeks spread successfully on the broad lowlands of much of western Asia under Alexander the Great. Equally he criticized Buckle for saying that the Indians are condemned to poverty by the physical laws of their climate, or that civilizations outside Europe were through the influence of 'nature' liable to possess imaginative faculties at the expense of reason. Nevertheless, such questions as the way of life of European settlers in the tropics and their powers of acclimatization are of considerable interest: there can be no harm in asking why the Japanese have never settled in large numbers in areas with cold winters, such as Manchuria, when the opportunities were available to them. Chisholm favoured caution in any effort to explain human life in environmental terms, and quotes with approval the statement of Jean Brunhes that 'every truth concerning the relations between natural surroundings and human activities can never be anything but approximate; to represent it as something more exact than that is to falsify it, is to become anti-scientific in the highest degree'. To search for a general law is a fascinating exercise, but Chisholm apparently thought it a wiser policy to look for empirical laws, which could be expressed by percentages or other numerical statements.

In its present development, geography owes much to the work done in the past hundred years. Apart from a vast accession of material, there has been a good deal of thought on its relevance and on the methods of study likely to produce good results. No subject can retain academic standing merely by announcing its methods of work, but only by carrying them out and letting the results speak for themselves, or, failing that, by asking the right questions even if no final answers can be given. Any study concerned with the distribution of population over the world, past, present and even possibly future, must be of relevance. There is no need to restrict investigation to areas at present occupied, as even areas of perma-nent ice and snow may become significant for temporary occupa-tion or for air routes, and with modern resources life at the south pole can be made quite tolerable, at least for a time. It is a not uncommon academic experience that someone begins to study something merely for interest, only to discover in the end that it becomes of great significance: it is also true that an idea may be put forward, widely accepted and applauded, but finally swept into oblivion. Another idea may attract little attention, but become popular—even fashionable—many years later: various examples are given in this book of such a nature. In cynical moments, the present author has thought that there are few new ideas in modern geography, but rather a number of old ones that have been put forward, forgotten, revived and in some cases used to good purpose. The modern interest in medical geography or in colonial study is following lines of inquiry suggested eighty or more years ago: modern physical geographers are considering matters that puzzled the pioneers of the American geological surveys in the days before W. M. Davis. Land use surveys of towns seem extremely modern until one remembers that some were done more than a hundred years ago. But here has been a vast increase in research work, and the present range of inquiry is well shown by the recent French and American reviews of recent publications.[6]

Much of the modern impetus was given by the 1914–18 war and the subsequent treaties. While it is clear that the re-drawing of the map of Europe was done partly with map evidence, including such distributions as those of nationality, language, and communications, the full story has not been told, though presumably the diaries and other private papers of Isaiah Bowman, when released, may prove informative.[7] Before the war, geography was already strongly established in France and Germany and a useful beginning had been made in Britain and America: speaking of Britain, H. J. Mackinder[8] said in 1935 that 'the half-dozen years before the Great War may perhaps be regarded as the divide between the dominance of the old and the oncoming in strength of the new kinds of geographical activity'. The splendid survey by W. L. G. Joerg[9] of European geography in 1922 showed that there was an excellent foundation for the advanced teaching of the subject in many universities: long taught for its practical value to students of commerce and for its general interest to others, including intending teachers, geography was now able to attract specialists in Britain as in France and Germany. In this critical phase of development, many British scholars turned to French writers for guidance on method, though before 1914 some such as Herbertson and A. G. Ogilvie had studied in German universities: after 1918, German geographers once more influenced British geographers, but in time the pulsating activity of American geography gave stimulus not only to British but to other European workers. At present the pioneer work of Russian geographers is being watched with interest by geographers in many countries: fortunately some of the abundant material they produce is available in translations. The Russians, too, have made translations of books published in other languages, in some cases without the knowledge of the authors concerned.

From its very nature, the subject must be international: for a time between the 1914–18 and 1939–45 wars some of its most distinguished exponents in Britain were so conscious of this that they used their lectures partly as a means of fostering support of the League of Nations and of liberal views on race relations. On a more formal basis, the congresses of the International Geographical Union every four years gather up some of the researches of the time and inaugurate commissions to deal with special problems, such as the world mapping on the 1:1,000,000 scale (p. 67) or a standardized representation of land use (p. 169–71).

Six Trends of Geography

Over the past hundred years, it would scarcely be possible to trace a series of consecutive phases without twisting the evidence into tortured generalizations of a chronological type; there have rather been six main trends of development which will now be considered. Of these the first, and most fundamental, is the acquisition of raw material by explorers and travellers and the less renowned, but generally far more thorough, field-workers of modern times: this may be called the *encyclopaedic* trend. Second, the need for efficient teaching of geography as part of a general education has long been recognized, especially by many of the Societies, and some part—but by no means all—of the modern advance indicates an *educational* trend. Third, the practical value of geography in assessing the potentialities of new lands and their problems led to a marked advance in commercial geography, and in time to wider studies, such as those of agricultural life, rainfall distribution and periodicity, and even conditions of health: this, in a general sense, may be termed the *colonial* trend. Fourth, efforts to trace a world pattern have always attracted some minds, and though this activity goes back into the early part of the nineteenth century, it was especially prominent in the early twentieth: it was a trend, or tendency, to *generalization*. Fifth, during and after the 1914–18 war, as in the works of some writers before it, the political implications of geographical distributions were increasingly realized, and consequently there was a definite *political* trend. Sixth, the natural recent development has been towards *specialization*.

The *encyclopaedic* trend was, in effect, one of exploration and recording of observations made with a varying degree of perspicacity. Some of the early compendiums of geography, such as Stanford's *Universal Geography*, are full of facts and essential in their day to later writers. Much of the knowledge of such areas as Japan and China came from the careful recording of journeys by travellers, without whose work geography in its modern form could not have existed at all. The attraction of reading about journeys into previously unknown areas has always been considerable, and the lectures of Livingstone, Stanley and similar travellers were thronged. Almost everyone is at some time thrilled by the account of an expedition to Everest, or a few months in the Antarctic, though the last area has now become a field of great scientific enterprise. Thirty years ago one listened to Antarctic lectures as adventures of fit young men with some interest in birds and glaciology: now one listens to accounts of the International Geophysical Year. Livingstone and other explorers were pathfinders:[10] he gives in his works much curious information, for example that 'intercourse with departed spirits' is considered witchcraft, or that 'the people seem to live in abundance. They have rice growing among the native corn. Only some of the women wear the rings in the lips. The rest are good looking. We never were visited by more mosquitoes than here'. Such an area, on the Zambezi, would now be visited by serious economic geographers, and perhaps also by social anthropologists.

By the eighteen-nineties, however, there were signs that some people were tiring of such stuff, and a group of French geographers founded the *Annales de géographie* with an academic purpose and the aim of avoiding 'nouvelles à sensation' (p. 62): even so, there are still many people who are impressed by tales of remote regions, though nowadays this kind of curiosity is likely to be best satisfied by underwater photography on the television screen. Opportunities of travel are always of value, but there is a quaint view among some people that one cannot be a proper geographer unless one has visited some area of great remoteness, and there are still people who proudly assert that they have been all the way to Timbuktu. In 1899, H. J. Mackinder made the first ascent of Mount Kenya, having decided that he must do something of the kind because 'most people would have no use for a geographer who was not an adventurer and explorer'.[11] But it is not by Mount Kenya that Mackinder is now remembered: the main phase of exploration has passed, and by 1914 the only major unexplored area outside the polar areas was Arabia.

In *education*, geography has often been advocated as a 'bridge' subject between science and the humanities: Thomas Arnold of Rugby[12] said in 1842 *[sic]* that 'a real knowledge of Geography embraces at once a knowledge of the earth and of the dwellings of man upon it; it stretches out one hand to history, and the other to geology and physiology: it is just that part of knowledge where the students of physical and of moral science meet together'. The early educational efforts of the Royal Geographical Society culminated in the famous Scott Keltie report which reviewed the position to the 1880's. Writing in 1913 on 'Thirty years progress in geographical education', Scott Keltie[13] noted that from 1905 the new secondary grammar schools had included geography as a major subject, that there were advances in the primary schools and in the training colleges for teachers. At this time, the real lack was suitable teaching in British universities comparable to that available in France and Germany. In Germany,[14] many early efforts were made but in 1893, it was urged at the annual Education Congress that geography should be taught in all classes of the gymnasia and similar institutions: the Germans at that time regarded the French as more enterprising, particularly from the 1870's (p. 46). Many of the pioneer modern geographers devoted much of their time to the encouragement of school-teaching, to extra-mural lecturing and to arranging summer schools: in Britain, for example, A. J. Herbertson, L. W. Lyde, H. J. Mackinder, H. R. Mill, M. I. Newbigin, E. G. R. Taylor and many more wrote textbooks, and even W. M. Davis in America did similar work. Academics of a later age generally leave the writing of textbooks to the school-teachers—though there are exceptions.

It is not possible, nor perhaps necessary, in this book to review the whole process of geographical education, but two observations must be made. The first is that in many schools in Britain—unfortunately by no means all—there has been a great recent advance in field-work tours, on some of which a great deal of original work has been done. The idea is not new—indeed there are records of such enterprises nearly eighty years ago: Scott Keltie,[15] for example, recorded in 1886 that the boys of Gordon's Hospital, Aberdeen, were 'taken out to the country, and in a simple, rough, but effective, and to them interesting and instructive way . . . taught to draw maps of a small area for themselves'. No doubt they enjoyed it much more than answering such questions as one in the Cambridge Junior examination of 1884, 'A person sets free seven carrier pigeons at Limerick to go to Belfast, Cork, Kildare, Kilkenny, Killarney, Tipperary and Waterford respectively. Draw seven lines from one point to show clearly the directions which would be taken by birds flying straight from Limerick to these towns. Which bird would have the longest, which the shortest, to fly?' What educational value memory work of such a character involves, is not clear. But field-work came in as part of the 'practical' training in geography, and some form of outdoor activity or visit can be excellently related to work in the classroom. A second observation on education[16] raises very wide problems indeed: in 1916, a report was issued after a conference of five British associations supported by school and other teachers, in classics, English, history, modern languages and geography. Recognizing that education must be provided for those of scientific and technical abilities, the conference urged that there should be a balance of interests, with provision for both humanistic and scientific studies, and that premature specialization in either field should be discouraged. Further, there should be an adequate training in the language and literature, the geography and history, of the pupil's own country and those beyond it; for the first object in all education is the training of human beings in mind and character as citizens of a free country. The modern advance of geography has been made possible by opportunities for specialization, but every subject can fructify another.

Colonial enterprise, a third main trend, has been prominent for the whole of the past hundred years: in recent years, it has perhaps become more prominent than ever before as various universities and university colleges in Africa have acquired well-staffed geography departments able to tackle local problems with sophisticated research methods. Interest in Africa arose particularly from the 1870's when various European nations were colonially minded: Americans at that time were more concerned to develop their own vast territories. Equally in Canada, Australia and New Zealand much of the geographical inquiry was related to settlement possibilities, though the Australians showed an interest in New Guinea

19

by the 1880's (p. 57). A large part of the former colonial world has now become self-governing, and one can only note in passing the stream of research work that has come from New Zealand and hope for more in due course from Australia: significant, too, has been the work done in South Africa. Canadian enterprise still includes a good deal of fundamental exploration, partly in climate and settlement possibilities, and has obvious points of comparison with similar work in the *taïga*, or northern coniferous forest, of the Soviet Union. The apparent amelioration of climate in arctic and subarctic latitudes, on which work has been done by H. W. Ahlmann for forty years, may have considerable relevance in such northern areas: the Russian and the Canadian geographer may still feel the satisfaction of penetrating *terra incognita*, or nearly so. It would be quite wrong to write as if the whole world were completely known already, for in fact large parts of it are as yet very little known, and so far subjected only to investigations of a reconnaissance character, which must be the prelude to more detailed work later. As noted on p. 45 the possibilities of medical geography were seen a long time ago, but so far they have only been partially explored.

A trend towards *generalization* dates mainly from the early years of this century, but was conceived long before in various efforts to show world distributions such as those of climatic types and vegetation. Probably it is essential to all geography teaching that there should be some idea of the world distribution of population, of structural belts with their apparently associated landforms, of climate and weather, of crops and vegetation. Certainly much of the appeal of British geographers such as Herbertson and Mackinder lay in their broad vision, and several writers have suggested a world division of climate: but new methods of investigating climate, such as those of C. W. Thornthwaite[17] may make acceptance of some standard schemes difficult. Equally, new methods of geomorphological analysis, especially those based on recognition of surfaces of erosion, may make schemes of physical regionalization based on structural elements highly questionable. And what is true of physical features is no less true of any human distribution: the aplomb with which Ellsworth Huntington explained so much in terms of climate and its fluctuations was gently rebuked by J. Scott Keltie[18] in 1913 when he wrote of him as 'one of the most active and original of our younger geographers, whose imagination may perhaps want a little of the taming that comes with years'. But did it? In later chapters of this book, several generalizations of the past have been noted, with some criticisms of them given at the time and later: at least one could argue that these generalizations have been a useful stimulus to further work.

As a term in a geographical sense, *political* has had many connotations. In some teaching it became the most deadly of all studies, in which unfortunate children learned off long lists of counties and the towns within them, possibly with a map. But it was thought after the 1914–18 war was over that there was a great political opportunity to redraw the map on lines of justice and equity, with plebiscites in certain contentious areas. To what extent the states formed from 1919 were on the lines advised by students of the map find field evidence still remains debateable, as many of the boundaries were intentionally strategic. Among geographers, Jovan Cvijić, the great Serb, undoubtedly did much to establish Yugoslavia, but the exact influence of the geographical advisers has never been assessed. At least the conferences making the new boundaries were adequately supplied with maps, and the various geographical journals of the day contained a number of interesting articles on the new frontiers. The new Europe became a subject of careful study, and notably so in some of the 'Géographie Universelle' volumes, such as that on Central Europe by E. de Martonne;[19] but within a short period after the war, the Germans began to mark their former frontiers again on all school and other maps. Bowman's *New World*[20] first appeared in 1921 and soon became popular through its vigour and originality.

One result of the war on the teaching of geography in universities was to restore the study of individual countries as units, particularly in Europe, even though systems of regional division might cut across international frontiers so that, for example, the Danube basin was shared by such potentially hostile states as Austria, Hungary, Czechoslovakia, Yugoslavia and Romania. If, as suggested on p. 94, Britain was ready for a new challenge in geography by 1914, the expansion of the post-war period was remarkable; and there was an equally effective expansion in the United States after 1918. Interest in the outer world increased considerably, especially when the possibilities of foreign travel were restored, though all through the inter-war period little was known of Russia, shut up in a protective silence with her successive five-year plans. But in many newly-defined countries the geographers were concerned particularly with their own resources and problems, including those of planning industry and agriculture: for example, some notable work was done in Germany on regional economic planning.

In time, what may be regarded as the present trend towards *specialization* emerged. Perhaps for this reason, when the new political geography edited by W. G. East and A. E. Moodie called *The Changing World* appeared in 1956, it was the work of twenty authors rather than one omnipresent Bowman. The range of material available on any aspect of geography has vastly increased, and consequently researchers and writers are generally working on a more restricted field than in earlier times. This tendency is seen perhaps in an extreme form in *American Geography—Inventory and Prospect*, which among types of geography includes settlement, urban, resources, marketing, recreational, manufacturing, medical, military; there appear to be twenty-seven varieties—a distinct change from the days when a university might include in its courses physical, human, regional with perhaps also economic and political. But this is an inevitable development: if one looks at the list of courses in an efficiently-run history honours school, one will find a surprisingly long list of subdivisions of history and of periods. And in the lecture-rooms the periods become shorter: indeed one known to the author covered only five years. Equally a geography honours school may now have a wide range of regional courses, few of which will cover areas of continental proportions, as formerly; and some courses in historical geography will cover only one period instead of wandering on from palaeolithic times to the twentieth century. Some look back nostalgically to the days of the 'broad general culture' provided, so it seemed, by geography thirty or forty years ago (and even more recently in places) while others, including the author, think that the future of geography must lie in closer study of a more restricted field, but by far more people than were available a generation ago. Of course there are many teachers in schools who find what they were taught in a university twenty or thirty years ago entirely satisfying still: they resemble the lady who proudly said that she admired her husband so much because he had never changed his mind about anything since he was eighteen.

Specialization and Generalization

One can only give examples of various studies which show this increasing specialization. On the physical side one could mention the fruitful results of long-continued observation of a few glaciers as indicative of current climatic changes (p. 114), or of the excellent work done at the Skalling laboratory in Denmark on coastal changes (p. 116) for many years: other examples include some of the team-work done, necessarily quickly but none the less efficiently, after any major disaster such as a large-scale flood (p. 117). Though the geomorphologist may habitually think in terms of a million-year time scale, he may see much within a lifetime, or even within a week. On the human side, one could instance much patient work such as detailed studies of particular towns, without which modern theories of urban geography could not have been made and by which the authenticity of such theories is continually threatened. Equally in any rural study, it is necessary to reinforce the statistical material by investigation of individual farms that are both the producing units and the homes of the people in the country-side. But perhaps of all changes the most radical is in regional geography: many now regard the traditional methods of proceeding from structure to physical features, climate, vegetation, natural resources, agriculture, industry, population distribution, types of settlement and the like as beyond the range of any-one: there is now a widespread tendency to focus such regional study around some theme such as the distribution of population and to seek influences in order that some explanation may be given.

At present the trend towards increasing specialization may make much geographical reading far less attractive, if more intellectually satisfying, than the broad generalizations of an earlier time. Many reviewers criticize works by geographers as lacking in apparent purpose, as being builders' yards from which others may choose their materials, as failing to substantiate some doctrine such as the necessity of preserving the countryside, of planning whole regions rather than single towns, of showing how a world food shortage may be averted, of advocating some industrial development in an area of declining economic power—the list is endless. To some it seems that no great planning problem can be solved except on a local basis, and that the necessity for such detailed study has been amply demonstrated by the rapid efforts of planners in Britain to make adequate surveys of the towns they propose to 'redevelop'. Much of this detailed work has raised new problems, both of research technique and inadequacy of data: for example in many countries the statistical material is deficient, yet who could say that the best possible geographical use is made of the statistical material available? Jean Gottmann[21] in 1950 said that 'the essential problems of our time . . . are permanent problems; those of planning and replanning the regions of the earth, of the compartments and the partitions chequering the continents— not only the surveying and description of all these, but also a permanent endeavour better to understand the principles and factors regulating the existing pattern in the organization of space!' Gottmann prefaces these remarks by saying that the geographers of the late nineteenth century—such as Ratzel, Vidal de la Blache, Mackinder and W. M. Davis—laid fine foundations by their preoccupation with 'broad and permanent problems of the world's partitioning' but it has become almost 'unscientific' to attempt again to solve the large issues, though this is an age of large issues the 'geographers have lost a great deal of the prestige and audience they had half a century ago'. Nevertheless, the success of such books as L. D. Stamp's *Undeveloped World*, or of various series of general summaries of geographical research in Britain and France, and the wide use of the Land Utilization Survey of Britain for planning purposes during the last twenty years show that some efforts have been appreciated.

One problem of which all practising geographers must be conscious is the speed of change in the modern world. The vast increase of the urban population in many countries, a varying rate of suburban spread, the reduction of numbers on the land combined with increasing mechanization in parts of the world, the development of new resources and the decline of some old-established manufacturing or mining areas, the growing pressure of population in some less well-endowed countries, the withdrawal of people from country-sides where returns are meagre, are among many changes that may be observed. Changes occur with varying speed from one country, or even one part of a country, to another, but one can only assume that even greater changes are impending. Much survives from former ages: much is changing before our eyes. In Britain the countryside has many villages and even farm-houses originally located in Norman times and still there, but in towns swift and dramatic changes may be seen: for example the 1959 report of the Ministry of Housing and Local Government[22] notes that the city of Birmingham has bought an area of 977 acres, at a cost including clearing, of over £18,000,000, for redevelopment. A visit to Birmingham will show the striking changes on the ground. Increasingly one turns to the city and town as the focus of modern life, never more accessible than now: increasingly one sees the attraction of industry and commerce to towns in a world where one clear desire, whether admitted or not, is a rise in the standard of living.

CHAPTER TWO

GEOGRAPHY FROM THE MID-NINETEENTH CENTURY

The mid-century challenge; the regional approach; some
systematic studies; the advance of cartography; the 1870's

As the nineteenth century rolled on into its later years, the need
for geographical study became clearer to many thinking
people. In 1858, for example, R. I. Murchison (1792–1871)[†]
in a long review of the year's events of geographical interest,[1]
showed that the discovery of new lands, and the penetration of
those already known, offered a vast challenge to enterprise and
scholarship. In Britain new government maps appeared annually,
and the Admiralty widened knowledge of the seas through its
charts and Pilot volumes: the deep-sea soundings were important
for the new Atlantic telegraph cables and the ocean oozes were
being examined and analysed by Professor T. H. Huxley (1825–92),
then of the Government School of Mines—later this led to the
study of oceanography in its own right. Explorers were penetrating
the Himalayas, and attention was turning to China where, Murchison argued, the real commercial artery was the Yangtze: could this
great river be opened to commerce, it would provide a route 3,000
miles long through a valley with 100 million inhabitants. Murchison thought[2] that the national judgement was unsound for, having
developed Hong Kong from 1842, 'we have attached too great an
importance to the territory around Canton, which is cut off from
the east, central and most populous portions of the empire . . . by
a chain of mountains at no great distance from the seaboard'.
Shortly afterwards, in 1862, Captain Blakiston[3] went 900 miles
farther up the Yangtze than any previous European traveller. On
north Australia, Murchison held the somewhat optimistic view that
it could be developed by convicts or perhaps rebellious Sepoys, for
whose control remoteness was an asset.

† This symbol indicates that a short biography is given in the Appendix
(pp. 303–25).

26

The Mid-Century Challenge

In all kinds of ways, at home and abroad, changes were seen. In 1860 Earl de Grey,[4] speaking of the Admiralty Surveys, noted the improvements made in the Tyne by the River Commissioners, and the removal of shoals in the Thames in Blackwall and Barking Reaches so that ships could reach the Pool of London at all states of the tide. In Canada,[5] the area drained by the Winnipeg, the Red River and the Saskatchewan was recently explored: 'with a vast area of fertile soil, and a climate favourable to the cultivation and growth of wheat; with lignite coal, iron-ore and common salt in abundance, a great future is in store for the Basin of Lake Winnipeg . . . as a British colony . . . instrumental in carrying British institutions, associations and civilization across the continent of America.' Australia too was changing[6] for within ten years its population had grown from 400,000 to a million. Trade with Siam was increasing; but of Japan[7] 'we knew nothing . . . beyond having seen a very few of its towns and a small extent of its highways. Not a man among us has acquired its language . . . there is no part of the world so little known to civilized Europeans'. Earl de Grey finished his address[8] by saying that he knew of 'no country in the world in which the results of geographical investigation are calculated to be of greater value than they are to England. With an empire that extends to every quarter of the globe, and embraces within its rule almost every variety of the human race, and with a commerce that fills every sea and occupies every port, the English have, perhaps, more to gain from the prosecution of geographical science than any other nation; and the researches of geographers are no less important to our statesmen, and our merchants, than to our men of science themselves'.

By 1860, there was already an impressive record of overseas discovery by British workers, with some distinguished contributions to cartography. The three earliest geographical societies of the nineteenth century, Paris 1821, Berlin 1828 and London 1830, were founded at a time of great geographical activity, but much had been done earlier—indeed, discovery and the recording of impressions of newly-visited areas is one of the oldest forms of human enterprise. It is intrinsically satisfying, and an inherent quality of young children who discover the world afresh and open up the world again to their parents, for there is often wisdom in the eye of a child. G. R. Crone and R. A. Skelton[9] have spoken of the great collections of travels in eighteenth-century England as manifestations of contemporary taste in 'an age of expansion in the arts of living': E. G. R. Taylor has written of Tudor and Stuart geographers, and of many more, not least the navigators of former times.[10] E. H. Bunbury† (1811–95) has written of ancient geographers[11] and in a fine summary of geography from 1500 to the nineteenth century J. N. L. Baker[12] said that 'the history of geography is long and honourable. No geographer need apologize for it or be ashamed of it'. He quotes from an address of H. R. Mill† (1861–1950) in 1901 the still relevant words:[13] 'We sometimes hear of the New Geography but . . . it is more profitable to consider the present position of Geography as the outcome of the thought and labours of an unbroken chain of workers, continuously modified by the growth of knowledge, yet old in aim, old even in the expression of the ideas that we are apt to consider most modern.'

Baker also effectively demolishes the idea that only people with a geographical training can write geography, for many others have made contributions of value, not least historians, classical scholars, divines, military strategists and others. These contributions were not necessarily written as geographies; in some cases the geography was introduced to explain things not otherwise explicable. George Adam Smith's *Historical Geography of the Holy Land*,[14] sheds light on the scriptures not discernible from other sources: the work of classical scholars like A. E. Zimmern and J. L. Myres is illumined by their close study of the Mediterranean and Near Eastern environment.[15] The debt of geography to other subjects, some of them its own children grown up to adult stature, is so frequently stressed that it is well to remember that other subjects owe much to geography.

Academically, however, geographers in the nineteenth century failed to deal adequately with the deluge of raw material. The world was opened up at a rate unprecedented in human history, and the geographical societies were well aware of the challenge of the times. C. R. Markham (1830–1916), in a review[16] of a century's geographical work from 1789—one of the most significant dates in European history—shows that the conquest of space was as important a phenomenon as the streaming flow of new ideas from the French Revolution onwards. In 1788, there was the first settlement at Botany Bay, and Flinders began to explore the shores of Australia: fifteen years later, Bass showed Tasmania to be an island.

In Africa, an association for discovery was formed in 1788 by Sir Joseph Banks and Major Rennell, and numerous journeys were made: the famous traveller, Mungo Park, reached Gambia in 1795. Major Rennell, already known for his maps of India, dated 1788, also produced a map of Africa in 1790. Arctic expeditions of this time date back at least to 1773, with the voyage of Captain Phipps, and in 1818, Captain John Ross rediscovered the great bay found by Baffin in 1616. In the Antarctic, the classic event was Ross's crossing of the Circle on January 1, 1841, to find an area of perpetual ice and the active volcano, Mt Erebus, 12,367 feet, in the journey to 78° 11′ S. Others had gone before him, but for a long period afterwards little was done to explore Antarctica, though in 1881–2 A. W. Greely† (1844–1935) went to 81° 44 N. on Ellesmere Island: his books on the Arctic were popular, and a lifetime of service included work on arid lands, meteorology and climatology.

All of the journeys and discoveries were consolidated by cartographers into permanent additions to knowledge. The first hydrographer was appointed by the British Admiralty in 1795 and knowledge of the seas and coasts were recorded with increasing efficiency: the British Admiralty still has the chart of the ship bearing Lord Amehurst and his retinue across the Gulf of Chihli in 1816 to Tientsin—it shows two lines of soundings, all that was known of a sea then so remote. But knowledge of many kinds came quickly, for some of the expeditions included 'naturalists' on their staffs, notably Charles Darwin (1809–82) and Joseph Hooker[17] (1817–1911), who succeeded his father to become the second director of Kew Gardens in 1865 and went on several expeditions, including the *Beagle* voyage of 1837–43. And the naval officers in charge of exploring ships included some men of great ability: one such was Rear-Admiral Sir Francis Beaufort†[18] (1774–1858). The son of an Irish clergyman who had produced a map of Ireland and a much-quoted estimate of its population in 1790 and 1791, he entered the Navy, fought in the Napoleonic wars, and from 1825–51 was Hydrographer. His widespread travels included the coasts of the West Indies, South America, the Falkland Isles, Australia, New Zealand, China, and the British Isles. He brought maritime soundings to great efficiency, but is also widely known for the Beaufort Scale, a method of measuring wind velocity, and a notation for recording weather.

Work of equal value was done by Vice-Admiral Robert Fitzroy†[19] (1805–65), who from 1828–30 and from 1831–6 in the *Beagle* surveyed the coasts of South America and the Falkland Isles, and in 1839 published the findings in three volumes, of which the third was by Charles Darwin. Later, from 1842–6 Darwin added three other volumes on the geology of the *Beagle* voyage, which on the return journey of more than a year included visits to the Galapagos archipelago, Tahiti, Australia, New Zealand, Tasmania, the Keeling Islands, Ascension, St Helena and the Cape of Good Hope. This voyage[20] was in 1865 said to have 'produced a harvest of fresh knowledge which from the combination of geographical, physical and natural history results, is unparalleled in this century . . . some of the best effects . . . have been from year to year coming forth in the writings of Charles Darwin'. Fitzroy, in his later years, induced the government to set up the meteorological office, where he established a system of storm warnings which developed into forecasts, especially at ports. Not all his enterprises were fortunate: from 1843–5 he was Governor of New Zealand, but he was recalled owing to disagreements on his policy of allocating land to the Maoris, and in his early visit to Tierra del Fuego, he brought back a family of 'wild Fuegians' to London, to improve and convert them, but the experiment was unsuccessful and he took them back on the *Beagle*, loaded with presents.

The Ordnance Survey of Great Britain[21] was begun in 1791 and ten years later the first sheets, on the 1 : 63,360 scale, appeared for Kent, part of Essex and London. The main purpose of the map was military, as for most similar surveys in other countries. But in Ireland the initial survey, on the 1 :10,560 scale, was intended for land valuation and the delimitation of townlands, the smallest and most ancient division the country possesses. From 1824, the work proceeded and the six inches to one mile maps appeared from 1833–46.[22] For many years the responsibility was carried by Thomas Colby† (1784–1852), an officer of the Royal Engineers who at times employed as many as 2,000 people.[23] Colby had three gifted junior officers, T. Drummond (1797–1840), T. A. Larcom† (1801–79) and J. E. Portlock† (1794–1864). Drummond[24] was largely responsible for the valuation survey; Larcom served on various government commissions, including the 1841 Census and organized a system of agricultural statistics from 1847; and Portlock,[25] having worked on the trigonometrical survey, became interested in the geological survey but ultimately returned to a military career. Apparently a man of iron, he lived under canvas at 2,000 feet in Co. Donegal in the depths of winter when working on his triangulation. Larcom[26] was responsible for the first memoir of the Ordnance Survey, on the twenty square miles of Co. Derry lying west of the Foyle, produced in 1837; unfortunately, it was the only memoir that ever appeared (p. 44). In the 1841 Census,[27] published in 1843, Larcom included a series of distribution maps showing the density of population, standard of housing, degree of literacy, and value of stock in relation to area, for all the baronies of Ireland. He also included a map of Dublin which gives valuable information on the town land-use of the time. And in 1838, the report of the Drummond Commission[28] on railways was illustrated by six maps, of which three were the works of H. D. Harness[29] (1804–83): these show the volume of traffic along main roads, canals and river navigations, the number of people using public conveyances, and the density of population per square mile—with the uninhabited areas excluded. There is also a map of Ireland on the scale of 1:253,440 and a geological map, attributed to Larcom.

This discussion of Irish maps may seem a digression from the main theme of the development of geography, but it is the subject for a brief sermon which will now follow. The real chance of the mid-nineteenth century was missed: an abundance of material was available for geographical study of the home environment, but it was neglected. The Ordnance Survey memoir scheme[30] of 1837 was killed by its very success, for the volume consists of 352 large pages, with many more of illustrations, including some in colour: it was a far more ambitious enterprise than the geological memoirs published later as descriptions of the 1:63,360 sheets. Had something simpler, shorter, less expensive and more selective been provided, there might have been a valuable survey of the entire country available for the mid-nineteenth century (see also p. 45). To a great extent the interest in the British Isles was satisfied by gazetteers, some of which, such as the topographical dictionaries of Samuel Lewis on England[31] (1831) and Ireland (1837), still provide interesting reading and, used with discrimination, remain valuable sources for students of historical geography. In Scotland[32] during the 1840's a statistical account of each parish was published, on lines developed from the earlier work of the 1790's. Equally interesting are many of the travel works of the time, and some much more serious accounts of various parts of the British Isles. The growth of towns, the world expansion of commerce, the increasing ease of travel, a growing interest in touring, all opened new opportunities and new problems. And all kinds of fascinating changes could be observed on the doorstep: perhaps because they were on the doorstep, they hardly seemed worth observing. The more remote and unusual an observation, the more it commanded attention; and this attitude still survives widely. Part of the trouble has been the lack of effective presentation of the geography of the homeland, as students of early textbooks will appreciate.

Nevertheless there were three hopeful signs of geographical development by the middle decades of the nineteenth century. First, there was a gradual development of what has been called general, or comparative geography, associated especially with the name of Carl Ritter†[33] (1779–1859), who stressed the idea of the interdependence of all phenomena on the earth's surface. As a young man, and throughout his life, he was deeply conscious of the relationship between geography and history, and his first book was a two-volume discussion of Europe (1804, 1807 'geographical—historical—statistical') with six maps, as a supplement.[34] 'This atlas,' says K. A. Sinnhuber,[35] 'in its sequence from mountain ranges, heights, wild plants, crops, animals to man shows how Ritter had grasped the interdependence of the main geographical phenomena.' Though it is untrue to speak of Ritter as 'the founder' of regional geography, his purpose was to produce a regional geography of the world by considering regions of relatively large size—a tendency widely prevalent well into the twentieth century. Far less attention has been given to the French writers,[36] both geographers and geologists, who from the late eighteenth century had worked out the details of natural regions in France, and made famous the names of the *pays* the country possesses.

The second great advance of the early and middle nineteenth century was in what is now called systematic geography; the *Cosmos*[37] (1845–62) of von Humboldt† (1769–1859) is frequently regarded as the main expression of this development but there were other, earlier works, including a French book on *Physiographie* (1836) by P. F. Eugène Cortambert† (1805–81), who was secretary of the Geographical Society of Paris.[38] The title is interesting as it anticipates by forty years Huxley's *Physiography* (1877), a term of which he regarded himself as the inventor.[39] In Britain, the *Physical Geography* of Mary Somerville† (1780–1872)[40] appeared in 1848, having been begun ten years earlier: in 1836, Mrs Somerville had published a work 'on the connexion of the physical sciences', which included a discussion of tides, currents, climate, plant geography and 'the infinite variety of organized beings that people the surface of the globe'.

The third great advance of the early nineteenth century was cartographical: reference has already been made (p. 30) to the growth of official surveys and several private firms added atlases, of which many were primarily concerned with political divisions but others also showed physical, including climatic and vegetational distributions. These firms also produced many local maps of distinction (pp. 227–33).

The Regional Approach

As knowledge of the world grew, the aim of some writers was to give an integrated picture of various regions in which all the inter-related phenomena, physical and human, were considered. Sharply criticized by many people at various times, this idea is far older than Ritter, and was advocated by Kant. Ritter[41] said that 'as chronology provides the famework into which the multiplicity of historical facts are ordered, the area (*Raum*) provides the skeleton for geography; both fields are concerned with integrated different kinds of phenomena together, each in its respective frame'. He was devoted to the idea of the unity of all nature, the inter-dependence of man and the earth: and so, too, was von Humboldt, though with different philosophical premises.[42] Ritter regarded all life as expression of a divine purpose, Humboldt[43] all 'humanity, without consideration of religion, nationality and colour, as one great closely related race, as one whole existing for the attainment of one purpose, the free development of inner powers'. To some extent, Humboldt's views of humanity were influenced by an aesthetic approach, derived in part from Goethe and various other intellectuals of his time: although his writing is mainly on non-human aspects, that was partly because his field investigations, especially in equatorial America, had taken him to areas 'where man and his products disappear, so to speak, in the midst of a wild and gigantic nature'.[44] His writings describe man, his culture and works as part of a general description and interpretation of nature.

The difference between the two writers—and both were by nature writers—lies primarily in their difference of aim. Ritter was actuated by a desire for a world view: von Humboldt concentrated largely, though not exclusively, on physical features, climate and vegetation, to such an extent that Hartshorne[45] has regarded him as the founder of plant geography and climatology. This may be too great a claim, but at least many would agree with Leighly[46] that Humboldt 'put a multitude of observations into a symmetrical theoretical system; he had the literary ability to present attractively both the observations and the system'. In the preface to the *Cosmos*, von Humboldt[47] noted that we are naturally led to consider each organism as part of the entire creation, and to recognize in the plant or the animal, not merely an isolated species, but a form linked to other forms either living or extinct. Admitting that much of the writing on the connection between natural phenomena and physical laws was ephemeral, or at least superseded by later discoveries, he hoped that 'an attempt to delineate nature in all its vivid animation and exalted grandeur, and to trace the stable amid the vacillating, ever-recurring alternation of physical metamorphoses, will not be wholly disregarded even at a future age'. Nor was it, for his integration of climate the vegetation became a main plank in later regional theory: Vidal de la Blache† (1845–1918), for example, the great French geographer, regarded the correlation of vegetation and climate as basic,[48] and notes that the 1837 Atlas of A. H. Berghaus† (1797–1884) brought out this relationship clearly for the first time. And more, the individual species always existed in association with others, all of which were the product of long evolution:[49] incidentally Humboldt's statement bears a strong resemblance to the idea of the 'survival of the fittest', which gradually became prevalent. Many of the detailed explanations of Humboldt are interesting:[50] he noted, for example, that the snow line was 3,000 feet higher on the Tibetan side of the Himalayas due, he suggested, to three factors, the radiation of heat from the neighbouring elevated plains, the purity of the atmosphere, and the infrequent formation of snow in a cold and very dry air. The basis of Humboldt's work was a five-year expedition to the Spanish possession in America, from 1799–1804, of which the results were published in twenty volumes from 1807–17, and a nine-month expedition in 1829 to central Asia, on which he published a book in 1843. His *Cosmos*,[51] published in five volumes from 1845–62 and considered by many at the time to be one of the greatest scientific works of the age, was partly based on his lectures at Berlin.

Perhaps no comparison is more odious than that between Humboldt and Ritter, though it is quite commonly made, sometimes crisply as in F. L. Kramer's comment 'von Humboldt has given us first-hand observations and measurements; Ritter methodological concepts'.[52] He adds that Ritter dealt with areas that he had never seen in his *Erdkunde*, and that 'the "founder" who attempted to show the way and who bade his pupils proceed from observation to observation appears to have been little of an observer himself. He uses the eyes of others much more than his own. Yet he lived in an era when science had just begun to find its eyes'.

The curious feature of Ritter's life was that though his extensive travels, apart from one short visit to Asia Minor, were all in Europe, only his early work (p. 32) was on Europe: the *Erdkunde*, published in nineteen volumes from 1817–59,* all dealt with Africa and Asia—the latter unfinished. His pupil, A. H. Guyot† (1807–84), argued that Ritter was so good a regional geographer because he acquired his regional training in Europe—the most varied of all continents.[53] But it is odd that he never even saw Palestine, on which he wrote so much, and Guyot says that Ritter commented, 'What new information could I derive from a visit to Palestine? I know every corner of it.' And doubtless he did from the journeys of others. His general method was probably influenced by the educational principles of Pestalozzi,[54] which involved three stages—first, the acquisition of the facts; second, the integration of all the information about a place for all times and third, the discernment of a general system, as it exists in nature. The title *Erdkunde*, or General Comparative Geography, excluded the merely descriptive, and belonged to the second, or integrative, phase. In fact, it proved to be too vast an undertaking for the third stage to be reached at all. Could it be said that Ritter tried to paint on too large a canvas, and that his apparent aim of covering the world was beyond the capacity of a single author?

To his contemporaries, Ritter introduced many stimulating ideas: writing in 1861, H. Bögekamp[55] lists many of these with considerable effect. Ritter stressed the idea of the land and the water hemispheres, the distinction between the rates of heating and cooling of land and water, the difference between the northern and southern hemispheres in their proportions of land and water. Then, too, there were differences between the continents—Africa had relatively the shortest and most regular of all coastlines and its interior least contact with the sea; Asia was better provided with

* The first volume appeared in 1817 and was soon out of print. The volume on Africa appeared in 1822, and the rest from 1832.

sea inlets, but the interior had little marine contact; and Europe was the most varied of all, with an ease of approach along its shore-line of comparatively great length. He had interesting theories on nations: each had its own individuality, as each was an aggregate of individuals, and each must have its own territory and localized position. From the contact between nations, he developed the idea of space relations: Palestine, for example, was from the begin-ning an isolated land, Israel an isolated people for thousands of years unintelligible to the world at large as all the great routes passed by it, and not through it. For this and other reasons, history and geography were closely interwoven: in Europe, for example, only in the Russian east was there uniformity of geographical features and uniformity of history, but in the west there was variety of environment and of history, and in the diverse south too history was rich, studded with the efforts and achievements of Egyptians, Hebrews, Phoenicians, Greeks, Romans, Gauls, Iberians, Carthaginians. Ritter advanced the theory of the north-west movement of civilization in Europe. Some of these ideas were developed in the book *Earth and Man*, by Arnold Guyot.[56]

It is easy to see how attractive these and other such ideas were to later generations of geographers. Some of them may seem obvious, and not all of them true: for example anyone who has read the history of Russia will hardly regard it as uniform but rather as tempestuous. The comment on the north-west movement of civilization in Europe is question-begging, though it appealed to the critical mind of Marion Newbigin† (1869–1934), as the first chapter of her *Mediterranean Lands* shows.[57] Some of the generali-zation on the continents have entered widely into the teaching of geography, at least in the more elementary phases, and H. J. Mackinder† (1861–1947) began his *Britain and the British Seas* with a comparison of the land and water hemispheres.[58] The intricacy of Europe as a continent is as well known to modern teachers as it was to Ritter and Guyot and perhaps for this reason many courses—both in school and university—reserve Europe for the later stages. 'Space relations' provide a stimulus to thought on the history of a nation, and the conception acquired meaning when treated by such teachers as P. M. Roxby† (1880–1947) of Liver-pool and, perhaps with less effect, by his disciples. Much that Ritter wrote has entered into geographical lore; some of his ideas were favoured for a time and then discarded, but revived as original later on. Many who advance ideas similar to those of

Ritter may be quite unaware of this origin, and it is by no means certain that all were purely the products of Ritter's fertile mind.

Of all criticisms of Ritter, and of many later geographers who viewed the world as their parish, none is more telling than the sheer impossibility of covering adequately so vast a programme of work. They looked too wide. One turns from this to an opposite extreme—the early efforts in France to give a firm basis to a study of the homeland. L. Gallois† (1857–1941) in 1908 showed that for many centuries names had been attached to regions in France which never corresponded to any political or administrative divisions—the *pays*.[59] Recognition of these was fostered by geologists from the mid-eighteenth century, especially in the Paris basin, and in 1808 C. Coquebert[60] (1755–1831) wrote a paper which distinguished such *pays* as the Beauce and the Gatinais: in 1817, his friend J. J. d'Omalius d'Halloy[61] wrote a paper in which the *pays* of the Tertiary formations around Paris were defined. These included the Beauce, a vast plateau of chalk, remarkably uniform, and mainly given to cereal-growing; the Brie, wet, with numerous ponds on the clays which at the surface alternated with calcareous sandstones; the Gatinais, low, wet, of poor soil and largely forest-covered. Various other writers drew attention to the relation between the geological structure and physical form, the plant life, agriculture and even the buildings made of local and traditional material.[62] Although the Paris Geographical Society favoured the idea of recognizing these regions, and as early as 1824 suggested that they should be widely studied, very little was done until after 1870. This was due in part to the idea that river basins could provide more effective units than the *pays*, which rarely corresponded to river basins; after 1870 the recognition of *pays* became favoured once more and every aspiring young geographer went out to write his regional study of a limited area. And from this new devotion to an old idea came one of the finest geography works ever written, P. Vidal de la Blache's *Tableau de la Géographie de la France*.[63]

Some Systematic Studies

Systematic studies in geography are interwoven with regional work. Ritter's work, though designedly regional, owes much of its significance to the general concepts that arose from it; and Humboldt's work, though more designedly systematic, derives much of its strength from its regional basis. It is unfortunate that some geographers of a later day have written as if there was some ranking of abilities between regional and systematic writers, in such phrases as 'the earlier geographers were regionalists as we of later age are systematists'. But every time has its arrogance, the most destructive of all academic vices. No writer was ever more immune from it than Mary Somerville, whose *Physical Geography* of 1848 was almost put in the fire as the *Cosmos* had begun to appear. [64] Being a woman of commendable sense, she listened to her husband, who advised her to consult Sir John Herschel, a distinguished scientist of the day: he urged publication and to him the work is dedicated. Her book received the warm praise of von Humboldt, an author sufficiently generous not to be actuated by jealousy. The book has fourteen chapters dealing with the land, five with the oceans, one with the atmosphere, ten with plant and animal geography, and the last with man. This arrangement shows both the progression from the physical to the human which is virtually standard practice and the concern of the times with the relationship between scientific phenomena over the world. Like later writers of successful works, Mrs. Somerville found the preparation of new editions a heavy task, as new material came in quickly. She brought out five editions of her work, the fifth in 1862, but the last two, dated 1870 and 1877, were the work of H. W. Bates (1825–92), of the Royal Geographical Society.[65] Although Mrs Somerville's work was successful, the trend of the times was for physical geography to become part of geology: J. N. L. Baker[66] showed that in the British Association, Section C, originally Geography and Geology, became Geology and Physical Geography in 1839, though from 1851 a separate section dealt with geography (and ethnography also until 1878). From this time, the main appeal of geography as such was exploration and travel, though in education there was some attention given to both physical and political geography. To a great extent, however, the idea of relationship was lost: readers of such a work as Samuel Haughton's *Six Lectures on Physical Geography*, 1880,[67] will observe that they consisted of some elementary geology, with a slight emphasis on physical features and on climate. And when in 1869 the Royal Geographical Society began its scheme of encouraging teaching in the public schools, medals were awarded for physical or political geography,[68] and the separate identity of the subject remained a theory question, for some regarded physical geography as geology.

Had geography any core, any central identity, or was it merely compacted of pickings from other subjects? This was an issue much alive a century ago: the historian, E. A. Freeman,† (1823–92)[69] for example, is quoted in the Scott Keltie Report of 1885 as saying that he did not see how geography could be recognized as an independent study in the universities, as 'on the one hand a large section belongs to the historian, and on the other side the geologist claims a large section of the field'. In the same report, there is an extract from the Cambridge Geology syllabus of a course labelled 'geological physics', which includes—in addition to palaeontology —climatology, meteorology, oceanography and much that would now be called geomorphology.[70] In Manchester, Professor Boyd Dawkins (1837–92) gave a course on physiography which included much the same material,[71] though with the significant addition of 'the distribution of man, and his advance in culture'. For the course the main textbooks were Huxley's *Physiography*, Geikie's *Physical Geography*, and Lyell's *Principles of Geology*, though Mrs Somerville's *Physical Geography* appears in that category familiar to all students, 'books of reference,' along with several more such as A. R. Wallace's *Geographical Distribution*, Wyville Thompson's *Depth of the Sea*, a work on anthropology and, most modestly at the end of the list, Dawkins's *Early Man in Britain*. The question of the placing of archaeology, anthropology and oceanography, was coming forward at this time.

One exponent of the unity of geography was William Hughes† (1817–76),[72] who wrote manuals for schools and delivered a number of lectures that were published. Two of these, given at Birkbeck and Bedford Colleges in London, dealt with the relation of geography to physical science and to history.[72] All geography, he argued, is basically physical, as all is Erdkunde, or earth knowledge. Geography, dealing with the world, rested on astronomy, and on geology, for 'everywhere one sees the influence of rock structures'. By observation of coasts, for example, one could see 'clear and recognizable changes', and the geologist, as Sir Roderick Murchison had said two years earlier, was 'but the physical geographer of former periods'.[73] Geography had its link with chemistry and physics in the study of such phenomena as mineral and thermal waters, or intermittent springs, and by its distributional approach geography could give life to meteorology: for example in 1869, the mean rainfall for the British Islands [*sic*] was 35 inches, but it ranged from 198 inches at Sty Head to less than 16 inches in East Lothian and Haddingtonshire.[74] Valuable also was his study of the distribution of flora and fauna: Hughes quoted as an advance the 'attributes and geographical habitat' of the tsetse fly, an insect whose distribution has been frequently mapped since.[75] Physical geography was a useful term, political was not, as he knew of 'no good reason why the description of peoples and industries, of manners and customs, or provinces and cities, should be called 'political' geography. It might with quite as much (perhaps more) propriety be entitled 'social' or 'moral'.[76] So far as the author knows, the last adjective has rarely if ever been used to define a type of geography, but 'social' now has an increasing vogue. On the historical aspects, Hughes comments[77] that 'every page of history bears evidence of the large extent to which the great actors in the drama of public life have been guided (often controlled) by the circumstances of surrounding locality . . . by . . . the geography, or topography, of a particular region . . . The statesman and the warrior are alike students of geography or find their schemes miscarry.' He was aware of the possibilities of geographical study of the Roman Empire, the Crusades, the Dutch landscape, the campaigns of Napoleon, and the potential fascination of studies of rivers such as the Rhine and the Thames. Unfortunately the greater part of his work was confined to school texts and a classical atlas[78] and his ideas were not worked out on an academic level.

Clearly the writers of systematic geography had to decide what it was, if indeed it existed at all. In America, Arnold Guyot[79] introduced some geographical ideas shortly after his arrival in 1848, on the invitation of Louis Agassiz (1807–73), the naturalist and glaciologist who consistently opposed Darwin's views. Guyot was a follower of Ritter, but had also attended the lectures of von Humboldt while a student in Germany. His first course of lectures in America was given at the Lowell Institute of Boston, on the relation between physical geography and the history of mankind, under a bequest to provide courses showing the harmony between natural science and revealed religion.[80] These lectures were published as *The Earth and Man*, which went through several editions: they show the influence of Humboldt and Ritter, and Guyot[81] states at the beginning that his work is to be no 'mere description', for geography 'should compare . . . should interpret . . . the *how* and the *wherefore* of the phenomena which it describes . . . endeavour to seize upon the mutual actions of the different portions of physical nature upon each other, of inorganic nature upon organized beings, and upon man in particular'. Guyot wrote various texts and finally a physical geography and produced a number of wall maps for American schools; but much of his research work was on the heights of mountains, measured partly by the barometer. He also set up several meteorological stations. An American worker of the same period was M. F. Maury† (1806–73), who was regarded by von Humboldt as the founder of oceanography, and was also distinguished as a meteorologist.

It is commonly said that Guyot had no successors, founded no school, and that American geography had to wait until W. M. Davis† (1850–1934) entered the field with his new geomorphology. But Guyot had a remarkable contemporary in George Perkins Marsh (1801–82), who has recently been the subject of an excellent biography by David Lowenthal.[82] Marsh had a career of immense variety and spent most of his later years as a diplomat in Europe: his *Man and Nature* appeared in 1864, having been begun in 1860 partly as 'a little volume showing that whereas Ritter and Guyot think that the earth made man, man in fact made the earth'.[83] Called by Lewis Mumford[84] 'the fountain-head of the conservation movement', Marsh's book showed that through ignorance, carelessness or greed, mankind lays waste the world at a rate that increases with his power to subjugate his environment. His views were far from those of the 'man and his conquest of nature' variety, for he shows the dangers of forest clearance: 'at one season, the earth parts with its warmth by radiation to an open sky—receives, at another, an immoderate heat from the unobstructed rays of the sun.'[85] He shows the dangers of soil erosion, and thought that forest removal changed the climate. Marsh wrote comprehensively of irrigation, and drew illustrations from his travels in Turkey, Egypt, Palestine, the Mediterranean countries and the Alps. The sub-title of his book was 'Physical Geography as modified by human action', and he summarized his views in the quotation given on the title-page, taken from a sermon of H. Bushnell:[86] 'Not all the winds, and storms, and earthquakes, and seas, and seasons of the world, have done so much to revolutionize the earth as MAN, the power of an endless life, has done since the day he came forth upon it, and received dominion over it.'

Marsh viewed geography as 'the science of the absolute and relative conditions of the earth's surface and of the ambient atmosphere (and) the investigation of the relations of action and reaction between man and the medium he inhabits'.[87] He looked to Humboldt rather than to Ritter and Guyot, and he regarded the material world as no evidence for God: to him spiritual religion must look elsewhere than the natural world for its evidences . . . 'it is a poor Divinity which rests its claims to godhead on the instincts of the beaver or the sagacity of the ant.'[88] Man was unlike all other natural phenomena, a free moral agent, presumably able to build or destroy civilization, and therefore man must keep dominion over nature by careful control and intelligent planning.

Marsh's work had various practical results in America. It stimulated the American Association for the Advancement of Science to submit a memorial on forests to Congress in 1873, of which the outcome was a national forestry commission,the designation of forest reserves and a national forest system in 1891.[89] In time this led to schemes of watershed protection, and eventually to a government programme for the conservation of all natural resources. On the planning of irrigation, Marsh also had considerable influence. He showed that it was not a simple matter, but might involve the building of reservoirs and the study of soil development, with dangers such as hardpan-formation, or the concentration of salts at the surface. In the west of the United States, the need was for a thorough survey of water resources from rain, melting snow and ground supplies, and of the potential use for agricultural, industrial and domestic purposes—in short, for a water code and public ownership. There was no question of merely digging a ditch and letting the water flow. Much of this material is inherent in *Man and Nature* but stated more specifically in *Irrigation, its evils, the remedies and the compensations*,[90] published in 1874: J. W. Powell† (1843–1918), in his *Report on the lands of the arid region of the United States* (1878) commended the work of Marsh and later, as chief of the Geological Survey, he incorporated a survey on the lines suggested. To a great extent Marsh's work was on applied geography and was remarkably fruitful; but like Guyot, he stands out as a somewhat isolated figure.

The Advance of Cartography

'When, in the far distant future,' said Sir Clements Markham to the Royal Geographical Society in 1880, 'the whole surface of the earth has been surveyed and mapped, the study of physical geography may be recommenced on a sound basis, and generalizations will become more accurate and will be founded on more correct and reliable data.'[91] He was speaking at the Jubilee celebrations and could point proudly to the advances of the past fifty years during which, for example, the whole interior of Australia and New Zealand had been discovered, explored and at least partially mapped: the maps of 1830 had been little more than inaccurate coastlines. Much had been done in Africa, regarded by Markham as 'a glorious field of generous rivalry among civilized Europeans'. The Indian survey was relatively advanced, and the heights of the world's loftiest mountains had been fixed between 1845 and 1850. Army officers had mapped the whole of Persia and Afghanistan, surveyed Mesopotamia, and explored the Pamir Steppe. Yet much remained to be done. Antarctica had been left unvisited since Ross's epic voyage of 1841-2 (p. 29), but in the Arctic the whole northern coast of Canada and much of the archipelago had been explored, though knowledge of Greenland was mainly confined to the coasts. The net result was that the atlases of 1880 gave a far fuller knowledge of the world than those of 1830, due not only to the discovery and mapping of new countries, but also to the great improvements in the methods of investigation, in the systematic arrangement of facts, in cartography, and in the construction and use of instruments. Allied to the exploratory surveys, there was the beginning of various national surveys, of which some examples were given on pp. 30-1.

Much was done by private firms. It is beyond the scope of this work to detail the long history of map-making in Britain, but the other books telling this story are themselves a contribution to geography.[93] Long before the official surveys, some excellent maps were produced, particularly of towns, and some of these have themselves become historical documents: as one of the Geographical Congresses urged (p. 67), it is imperative that all maps should be dated. Many of the maps in journals were produced by the private firms: of John Arrowsmith† (1790–1873), for example, it was said that there was 'scarcely a map now extant that does not bear the impress of his patient toil in the collection and arrangement of materials, often most crude and discordant, to show the progress of discovery'.[94] In 1832 he had prepared a map for the Royal Geographical Society showing the journeys of Captain Sturt 1795–1869), who traced the courses of the Murrumbidgee and Murray rivers to the sea, and partly decided the current controversy on the interior of Australia—was it a vast inland sea or a burning desert?

German cartographers made notable contributions: for many

years the von Sydow maps and atlases, with some wall-maps, were widely used. Emil von Sydow[95] (1812–73) was an army officer who retired in 1855, and established himself in Gotha, where Bernard Perthes ran his famous publishing firm: by 1867, von Sydow's school atlas was already in its twentieth edition. Meanwhile, in 1860, von Sydow had returned to the Army and he was in charge of the map supply for the Franco-Prussian war of 1870–1. The Perthes firm were responsible for the publication of Stieler's atlas, which was largely the work of A. H. Petermann†[96] (1822–78), who became director of the Gotha Geographical Institute in 1854. His six-sheet map of the United States, published in 1875, was for many years a standard reference source. Petermann had been a student at the Geographical Art-School founded by Berghaus at Potsdam, and made a map showing the mountain system of Asia for von Humboldt. For some years Petermann worked in London, and produced some interesting population maps, including one for the British Isles with a stipple shading for densities, and town populations in coloured circles. This[97] was published in 1849, and makes the curious mistake of assessing Ireland's population at 8,495,589 on the assumption that it had continuously increased from the 8,175,124 of 1841: in fact there was a catastrophic fall through famine deaths and emigration after 1845. Later work by Petermann[98] includes population maps of England and Wales, and of Scotland, in the 1851 Census, both of which have a very delicate shading to show population density; and in 1852 a map showing the distribution of cholera in 1832, with a detailed inset for London and comments on the incidence of the disease in his *Statistical Notes* (see also pp. 229, 231–2).

Petermann spent two years, 1845–7, with W. & A. K. Johnston in Edinburgh,[99] where he was employed on the preparation of the *Physical Atlas*. This was derived partly from Berghaus's famous atlas (published 1837–48), long known to students from its commendation by Paul Vidal de la Blache, who said that it was a landmark in geographical studies as it brought out in map form the relation between climate and vegetation. The Johnston firm was run by William and Alexander Keith Johnston† (1804–71) from 1825, and their first major work was *A Traveller's Guide*, 1830, followed by the *National Atlas* in 1843. Meanwhile A. K. Johnston travelled in Germany and Austria in 1842 and bought from Berghaus the copyright of some of his maps. In 1845 he went to Paris and met Humboldt, who gave valuable advice on the physical map, and regarded it as an improvement on the work of Berghaus. The first edition contained thirty maps, half of which were partly derived from Berghaus. The success of Johnston's *Physical Atlas of Natural Phenomena*, 1848, did much to establish the term physical geography at this time. Later, Johnston produced a globe, thirty inches in diameter, a dictionary of geography—in effect a gazetteer—several other atlases called general, classical, physical, astronomical, and one elementary atlas. From 1855, he began the *Royal Atlas*, of which every sheet was seen by the Prince Consort before publication. One interesting effort, in 1852, was a large-scale drawing of a chart of the geographical distribution of health and disease, given also on a reduced scale in the second edition of his *Physical Atlas* (1856): medical geography is by no means a new idea. E. W. Gilbert,[100] commenting on the association of Johnston, Berghaus and Petermann, notes that 'the first maps of the geographical distribution of diseases to appear in any German atlas' were published in 1847 as part of the Berghaus's *Physikalischer Atlas*.

Many other cartographers made great contributions to geography, not least the Bartholomew family of Edinburgh, of whom in 1880 J. G. Bartholomew† (1860–1920) introduced layer colouring on maps and produced numerous atlases (see also pp. 238, 240). Without the advance in cartography, the modern advance in geography would not have been possible. Even more important, many workers regarded cartography as basic to any social advance: Colby,[101] for example, of the Irish Survey (p. 30) regarded his work as contributory to any national improvement, and as the groundwork for historical, antiquarian, natural history, geological and statistical surveys. He took pride in the fine statistical work of the Irish Census, which could not have been done without the mapping of townlands, parishes and boundaries; illustrated by maps, the nineteenth-century Irish censuses did much to reveal the distribution of the country's problem areas. In the preface to the 1881 Census,[102] it was said that 'the uses of maps and diagrams in illustration of statistics are now so universally known that it is unnecessary to refer to them here'. On p. 31 reference was made to the Templemore Memoir: in 1844, a report of the Commissioners[103] appointed to inquire into its cost was published, and Sir Robert Peel noted that it included a vast amount of detail 'on many points of merely local or temporary interest'. The evidence given to the commissioners included suggestions for county memoirs, for thorough local social surveys, and for improvements in the Ordnance Survey—especially by contouring. One is tempted to quote many witnesses, but one must suffice—Sir Robert Kane (1809–90) author of *The Industrial Resources of Ireland*, 1844. He wished to have local memoirs showing for each the 'physical character' and climate, fuel resources, soils, agriculture, size of farms, minerals, fishery resources, and 'industrial intercourse', or trade at home and abroad. This, he urged, could well be collected by the survey officers: it would make possible the intelligent use of capital and labour; would shed light on the relation of plants, animals and soils; would reveal the distribution of mineral ores and waters and types of soil (which could be analysed—as he showed practically). Not only Kane, but many more witnesses of 1844, had similar ideas, and, as noted on pp. 63–4, so had H. R. Mill two generations later.

The 1870's

Of European countries, none made a greater advance in geography than France during this ten years: this is generally attributed to the national defeat in 1870–1[104] which stimulated people to look to Africa and other foreign fields as ground for new commercial opportunities and for the spread of French civilization. Goethe had said 'ce qui caracterise les français, ce n'est pas leur politesse, leur esprit, leur grâce, leur clarté, c'est leur ignorance en géographie'.[105] But there had been advances in education: from 1857 the subject had been taught in the primary schools, partly with a view of developing the power of observation, strongly urged by Pestalozzi and other educational reformers and partly 'à éveiller en eux l'amour du lieu natal'. The state syllabus of 1857 was based on the principle of working from the child's immediate circumstances to the canton, the arrondissement, the département, then to France, other countries and the world. Teaching in the lycées and collèges developed later, mainly from the 1870's, but in 1870 geography was taught only in the universities of Paris and Nancy, where Vidal de la Blache was already well known. As noted on p. 52, several new geographical societies were founded at this time in the provinces, and there were various writers such as Vivien de St Martin[106] (1802–96) and E. Reclus† (1830–1905), whose work was widely read. St Martin was from 1840 secretary of the Paris Geographical Society, and from 1862–76 published the 'Année géographique', a survey of the year's progress: he wrote voluminously, and thought that the strength of geography in Germany was largely due to the excellent manuals and atlases available. Reclus,[107] a student of Ritter, published *La Terre*, a physical geography, and the *Nouvelle Géographie Universelle*, a regional survey of the world in nineteen volumes, from 1875-94.

The outlook of the times was reflected in a survey of 1876 by Charles Maunoir,[108] secretary-general of the central commission for geography. He begins triumphantly by saying that the term *terra incognita* might become obsolete, and mentions the publication by the Paris Geographical Society of 'Instructions aux voyageurs', which were to serve much the same purpose as the famous 'Hints' of the Royal Geographical Society (p. 55). The voyages of A. E. Nordenskjöld† (1832-1901) in the Arctic were being followed with interest, and in North America the long peninsular finger of California was being explored. Efforts were being made to provide a map of Brazil: 'how unfortunate that this vast country . . . has not seen as much exploration as the United States.' Having reviewed other new discoveries, the writer notes that Africa should be the concern of the immediate future: Europe's great neighbour, it was being opened to civilization by travellers annually more prepared for eventualities, due to accumulated experience, though courage and audacity was still needed. Most of the journal space of the Paris Society was occupied by exploration, though in 1876 there was also an article on the Basques, somewhat apologetically introduced by the comment[109] that it was 'naturel de rechercher sur notre propre sol les questions qui s'offrent à nos études'. But the practical uses of geography were coming to the fore, and of the French journals the most frank was the Bulletin of the Marseilles Society,[110] which in 1877 said that it did not pretend to be a learned journal: rather was it 'un oeuvre de vulgarization', concerned with the use of geography for shipping, commerce, industry, agriculture, statistics and political economy—what in one university was called 'realistic research'.

In other countries there was a varying amount of progress during the 1870's, with Germany well advanced in Europe. Although in America Guyot's book had been well received, and translated into French, yet its influence quickly faded. In Britain the Royal Geographical Society abandoned its scheme for giving medals to schoolboys, as it had been only partially successful (p. 57); but the traditional interest in exploration and commercial expansion remained, combined with the hope of advance in education. It was through the appeal of the wider world that the next surge of interest in geography arose in the 1880's—Africa, the New World, a partially known Asia, not to mention the polar seas, all attracted vast interest as nations in the surge of industrialization sought new economic, and sometimes political, conquests beyond their home territory.

CHAPTER THREE

EXPLORATION AND EDUCATION:
THE WORK OF THE SOCIETIES FROM 1820 TO 1900

The earlier foundations; Geographical Societies after 1880;
the academic possibilities

GEOGRAPHICAL SOCIETIES fall into no clear pattern. At one end of the scale they consist of a group of people assembling to hear lectures with such titles as 'Sidelights on sunny Spain' or 'With a caravan in Southern Ireland', illustrated with films or colour transparencies; at the other they consist of earnest professionals assembling at infrequent intervals to hear papers of so esoteric a character that their authors appear almost as puzzled as the audiences; but between these two absurd extremes much excellent work has been done. Many societies have failed to adjust their work to the needs of the times; some have perished or retired into semi-oblivion with an annual dinner and possibly a lecture also, while others have endeavoured to meet some current wave of opinion such as nationalism or colonial expansion. In some societies the original objectives, such as the provision of practical information for merchants, have proved to be beyond their resources, and in others there has been a drastic change of policy, such as that in the National Geographic Society of America which transformed its journal into a popular form after a period of austerely academic work. Though all societies depend on the support of their members, a number acquire some form of subsidy from governments, local authorities, or universities, yet the great growth of geographical societies in the nineteenth century was due primarily to public interest.

The Earlier Foundations

Who supports geographical societies? In an anonymous note on the jubilee of the Russian Geographical Society,[1] founded in 1845, it was reported that the initial meetings in St Petersburg were attended by four groups of people—navigators (including Wrangel), the 'academic' circle of the 'naturalists' (including

49

Köppen the statistician), officers of the General Staff, and those anxious to promote Russian science (including geographers such as Perovsky, later famous for expeditions in Central Asia). The society worked in four sections, mathematical, physical, ethnographical and statistical. In its first fifty years there were numerous scientific expeditions within Russian territories and beyond them, including Polar regions. The desire to abolish serfdom, prominent among the enlightened in 1844, led to the investigation of such problems as the customary law, the organization of landed property and the judicial customs prevalent both in Russia proper and among the mountaineers of Caucasia, the Moslem populations of Turkestan, or the natives of Siberia. Though never having a large membership, the Imperial Russian Geographical Society acquired respect for its contribution to economic and historical studies and became influential in public life. From 1846 it published memoirs from its headquarters in St Petersburg, but later it had a number of sections, all of which published one or more periodicals: these included Tiflis (1852), Irkutsk (1856), Orenburg (1870), Omsk (1879) and Vladivostok (1888). Apart from its valuable scientific work, which included the founding of meteorological stations, one of its early aims was to prevent the wholesale destruction of social institutions by the state among the diverse groups forming Russia. No doubt the society's conception of geography was wider than that commonly held today, but the relevance of its work was clear. As in all societies, the quality of its membership was more important than the quantity (though quantity pays subscriptions); it was aided by government grants.

Although there were societies founded before the nineteenth century with geographical aims, none survived in Britain and little was done by the Royal Society. C. R. Markham[2] showed that from 1665–1848 only 77 of its 5,336 papers were on geography and several of these were on marginal subjects such as the observation of the track of Venus at far distant points on the earth's surface. But the British Association for the Advancement of Science has, from its foundation in 1831, always welcomed geographical work, and has developed its own Section E since 1851: until 1914 the work of the section was organized by the Royal Geographical Society.[3] The world's first specialist geographical society, founded at Paris in 1821, was followed by the Berlin Society in 1828 and the Royal Geographical in London, 1830: other societies included the Mexican, 1833, Frankfurt-on-Main, 1836, the Brazilian Instituto Historico e Geografico in 1838, the Russian in 1845 and the American in 1852.[4] In 1832 officers of the East Indian Company's Navy formed a geographical society at Bombay for Asiatic exploration, which became a branch of the Royal Geographical Society in 1833 but returned to independence in 1837 and survived to 1873, when it was merged with the Bombay branch of the Royal Asiatic Society. Although the London society provided for the affiliation of branches overseas,[5] no branches were founded, and in the 1880's a request from the Tyneside and Manchester Geographical Societies for representation on the council of the Royal Geographical Society was not granted. Many of the societies had difficult years, and even the Paris society, though active in promoting voyages of discovery and in publishing bulletins that included numerous maps and travel wonders, hardly increased in numbers from its initial 217 until 1860, when it grew steadily to 600 in 1870[6]. But after the national disaster of the Franco-Prussian war, a number of newspapers as well as the government's *Journal Officiel* began to publish geographical articles dealing with the outside world: mortified at home, France began to look wide. Later so distinguished for work on the homeland, the French at this stage became colonially minded, and geographical societies multiplied. The appeal made by the Paris society for popular support was based on two arguments: first, that the superior geographical knowledge of the German armies had led to their victory ('it was the schoolmaster who triumphed at Sedan'), and second, that the triumph of civilization over barbarism was 'an essential condition of the prosperity of nations'.

The need to spread civilizing influence over the more barbarous parts of the world was widely accepted by public opinion at this time. Geographical societies not only satisfied a natural curiosity on the more savage aspects of nature and society, but also cast a shrewd glance at the eventual possibilities of trade and colonial expansion. Christian missionary enterprise sought new fields to conquer, notably from Great Britain where the work of Livingstone (1813–73) was followed with breathless interest. In France some writers looked to North Africa not only as a base for expansion, but also as an area open to the settlement of French people, especially those who no longer wished to stay in a German-ruled Alsace. The Society of Commercial Geography,[7] founded at Paris in 1876 after working for three years as a commission of the Geographical Society, had as its aims the world-wide development of French commerce, the advancement of knowledge of commercial geography, the stimulation of voyages that might open routes for French trade and industry, the study of natural resources and manufacturing processes, and the study of colonization and emigration. From 1879 a bulletin was issued. Meanwhile, in 1873, a strong society had been founded in Lyons[8] with comparable aims and an educational purpose met by providing courses on commercial geography in schools. In the following year a group of merchants and shipowners in Bordeaux founded a similar society which became the centre of a group of societies in towns of the south-west and had 1,300 members by 1881.

Partly through official encouragement, the French societies rapidly acquired kudos, and they provided a delegation at the International Commission for the Exploration and Civilization of Africa, organized by King Leopold II of the Belgians in 1876. In the previous year the Paris society had organized the second International Geographical Congress with success, and impetus was also given by an exhibition of colonial produce in 1874 and the Paris Exposition of 1878.[9] In 1875 a section dealing with scientific and practical geography was added to the Natural History Society at Toulouse and in 1876 the Marseilles Society of Geography and Colonial Studies was founded: almost at once a quarterly bulletin appeared and the society sent explorers to discover the sources of the Niger in the combined interests of geography, of commerce and of France. From 1877 the *Revue de Géographie*[10] frankly advocated colonialism: indeed when France acquired Tunis in 1881, the journal claimed some credit. The year 1878 saw the foundation of three societies, at Oran, Montpellier and Rochefort. Of these the Montpellier society was concerned mainly with local geography, but the Rochefort society attracted many men with a naval background and colonial officials, and paid special attention to Indo-China. Two more societies were founded in 1879 at Nancy and at Rouen; the Société Géographique de l'Est at Nancy, so near the new frontier, was concerned not only with 'backward countries' but also with local geography. In 1880 the Union Géographique du Nord, from a headquarters at Douai, included thirteen autonomous societies in neighbouring towns.[11]

Geographical societies were founded all over the world at this time. In South America a society at Pernambuco began publication in 1863 and two more journals appeared at Buenos Aires in 1879 and 1881. The Egyptian Société Khédiviale launched its first publication at Cairo in 1875, and many more were added during the next ten years.[12] By 1885 the world had 94 geographical societies, with over 50,000 members, of which 80 were in Europe, including 26 in France (18,000 members, 34 journals) and 24 in Germany (9,300 members, 28 journals).[13] Most societies published journals or at least some annual or quarterly bulletin, but not all the journals were controlled by societies. Even Switzerland had six societies, but Britain and its colonies only five, partly due to the dominance of the Royal Geographical Society. By 1881 only Turkey, Greece, Serbia and Norway in Europe lacked such societies: in 1865, there had been only sixteen in the world. After 1885 there was no substantial increase in the total number of geographical societies, though in 1896 Wagner's Yearbook reported 107, with 38 branches, in 22 countries: 153 serial publications appeared of which 125 were published by societies, including 48 in French 42 in German and 15 in English.[14]

The American Geographical Society, founded in 1851, had much in common with European societies. Like them, it was interested in Africa and thrilled by the activities of Livingstone, but hardly colonially-minded as its interest in territorial expansion was abundantly met nearer home. At one of its early meetings in 1851, a trans-continental railway was advocated, and as late as 1859 it was proposed that there should be expeditions into the 'immense and little-known region' between St Paul (Minnesota) and western British Columbia, now fertile farmland with such towns as Edmonton and Winnipeg.[15] Interest in the Arctic was natural enough, and help was given to Dr E. K. Kane (1820–57) to organize an expedition to test his theory of an Arctic iceless sea. Rescued by a relief expedition in 1855 after two winters in Kane basin to the north-west of Greenland, Kane published two books on his experiences, which sold widely.[16] Special attention was given to South America and a paper given by E. A. Hopkins, American consul to Paraguay, in which he made a bitter attack on British and French trade on the river Plate, drew forth a sharp protest from Sir Roderick Murchison at the Royal Geographical Society.[17] The American society, as a practical expression of the Monroe doctrine, maintained its interest in Paraguay during the 1850's. From its beginnings, the society was interested in exploration, and the real lure for members was remote and unfamiliar places, such as the Arctic, Africa, the South Seas, and also the American Far West. During the 1860's and 1870's a vast amount of geographical work was done by government agencies, much of it so hidden in files and reports that one speaker commented 'the difficulty of ascertaining what there is . . . is only surpassed by the difficulty of knowing how to get at it.'[18] Many research workers have had similar feelings. An interesting development in 1872 was the Congress bill to make the area of geysers and hot springs near the sources of the Yellowstone and Fire Hole rivers, then newly discovered, into a National Park.[19] At all times the Society showed a vast interest in its homeland, itself the object of perpetual discovery and rediscovery. In the 1870 Census volumes some distribution maps were included and received with guarded approval by the Society.[20] Though not blessed by an unchequered history, the American society preserved its wide interests and became one of the greatest in the world.

In London, the Royal Geographical Society, founded in 1830, began with some 460 members, including a number of distinguished travellers. Among the founders[21] were Robert Brown (1773–1858), one of the first explorers of the Australian and the keeper of Botany at the British Museum; T. F. Colby whose abiding monument is the first six inches to one mile survey of Ireland; Roderick I. Murchison who, having retired from the Army in disgust at the failure of his regiment to fight at Waterloo, ultimately became a geologist of some fame; and the Hon. Mountstuart Elphinstone (1779–1859), at one time Resident in Poona and later Governor of Bombay, who spent the years of his retirement from 1830 working at his history of the Mogul rule in India. From its early years, the society published reviews of the geographical work done during the previous twelve months, given as presidential addresses: most of these are fascinating to read now. At all times the Royal Geographical Society was far less commercially conscious than many of the societies already named, and at no time could it be accused of favouring the 'Brazil-where-the-nuts-come-from' type of geography. To a very large extent its work has been the encouragement of exploration, and as early as 1831 it absorbed the African Association founded in 1788 and the Palestine Association founded in 1804 for the exploration of Syria and the Holy Land.[22] From 1879 to 1926 practical instruction was given in surveying to intending explorers, with classes also in geology, botany, meteorology, zoology, anthropology and even photography, as well as in geography.[23] And many aspects of exploration are covered in the well-known *Hints to Travellers* which first appeared in 1861 and has gone through many editions since. The information given is vast, and covers all the subjects on which practical instruction was given at the society, with much more that might be useful, such as the best methods of enlisting the friendly co-operation of yaks: in the eleventh edition, we are informed that 'The yak does not like the scent of a European, who should not stand near during loading'. A. R. Hinks† (1873–1945) revised the *Hints* and acted for many years as secretary to the Permanent Committee on Geographical Names. The Society has made a great contribution to geography, but too much of its journal space has been filled with such comments as these (chosen at random) from quite recent numbers: 'For about an hour and a half we climbed steadily upwards, following an excellent and well-graded path high up along the side of a deep valley. At the top of the pass we were met by a fresh relay of bearers and a police sergeant from Toa, where we were to halt at noon.' Or again—'Alcohol when at rest was useful as a stimulant or appetizer, but on the move was definitely harmful.'

The Royal Geographical Society has often been accused of obscurantism and traditionalism, and particularly of favouring exploration and mathematical geography at the expense of new thought in the subject. In the 1880's, however, its vision was wide. At that time it sponsored the work of J. S. Keltie† (1840–1927), who investigated the subject's educational possibilities (pp. 18–19), and it gave both good advice and financial help to universities beginning geographical teaching, long after the subject was well established in French and German universities. Confronted with a world rapidly being made known by explorers it gave its medals not only to travellers but also in 1869 to Mrs Mary Somerville,[24] whose books included a remarkable physical geography published in 1848 (see p. 38). And in 1845 the Society had honoured Carl Ritter, who was described as 'the first geographer of the time'. Ritter's teleological views have often been misrepresented, but in 1861, W. L. Gage[25] said that 'Ritter nowhere makes man subject, in silent, unresisting dumbness, to the influences of nature; he allows for results which issue from the mutual conflict of geographical conditions and the freedom of man's will'. The Society also honoured Baron von Richthofen† (1833–1905) in 1878 and some eminent cartographers, including John Arrowsmith[26] who produced the first edition of *The London Atlas* in 1834 and made substantial contributions to the mapping of areas newly discovered.

Much of its enterprise was due to men like D. W. Freshfield† (1845–1934), who devoted most of his time to societies, including the Alpine Club, and was an accomplished mountaineer: he was deeply concerned with the educational possibilities of geography.

Geographical Societies after 1880

No geographical society of the late nineteenth century existed in isolation and there is considerable evidence that through occasional congresses as well as by individual contacts, geographers in various countries watched each other's work with friendly interest. They may indeed have been almost too eager to take in one another's washing, even to the extent of reprinting translations of articles, of which, for example, there were a considerable number in earlier issues of the *Scottish Geographical Magazine* and the *Geographical Teacher*: nowadays some societies and journals apparently feel that they have their own washing machine. In the last twenty years of the nineteenth century the rate of discovery was breath-taking: in 1883, the R.G.S. president[27] said that 'the study of practical and scientific geography is being prosecuted with an ardour and energy never exceeded in any age in the world' and in 1884[28] he added that 'the area of the unknown is . . . steadily diminishing, under the attacks of travellers of various European nationalities, some of them with scanty means but an abundance of scientific enthusiasm, others supported by powerful associations but with objects not purely geographical'. In the 1883 annual address, for example, Lord Aberdare spoke of nine observatories recently established on the Arctic, including those of Britain on Great Slave Lake, of Austria–Hungary on Jan Mayen, of Norway in Lapland, of Sweden on Spitzbergen and of Russia on Novaya Zemlya. In Greenland Baron A. E. Nordenskjöld disproved his theory that its central valleys had a comparatively warm and dry climate by travelling inland for 230 miles to find land 6,000–7,000 feet high covered with ice. In Africa the Masai people, 'so little known . . . an object of terror,' were visited and there were various travellers on the Congo. In Asia, the triangulation of India was completed, and some details were given of the native states of the Malay peninsula, and also of Afghanistan and Persia. In the East Indies, a good deal of knowledge was added, largely by pioneers of the North Borneo Company who explored the north part of the island. New Guinea was still little known as the missionaries on the coast had never ventured more than twenty to thirty miles inland, still less to cross it; in the following year one of the missionaries went farther and found a cannibal tribe. Various expeditions visited parts of Central and South America from Guatemala to Tierra del Fuego. And in every year the Navy reported the advance of its work in surveying coastal and river channels: the famous Pilot volumes, graced by a tradition of good clear writing, increased in number.

Unfortunately the vast expansion of exploration was not supported by adequate teaching in schools and universities. In 1885, the Royal Geographical Society[29] abandoned its scheme of presenting medals to schoolboys for good work, as out of sixty-two awarded since 1869, half had gone to Dulwich or Liverpool College and only fourteen other schools had submitted candidates. Scott Keltie was appointed to investigate the teaching of geography in universities and schools and his report of 1886 emphasized the poor quality of the teaching, and its weak, almost negligible representation in the British universities compared with those of France and Germany. With justice, Lord Aberdare[30] could say 'Geography does not pay. It is not recognized at the universities . . . it does not find a real place in any of their examinations; while in the Army and Navy examinations it is at a discount . . . merely left to crammers'. Yet the maps, especially political maps, were changing rapidly: Germany was annexing land under the control of the Society for German Colonization in East Africa, and in New Guinea, which was partitioned with Great Britain: Britain established the Bechuanaland protectorate and the protectorate on the Niger. Expeditions were numerous, and the new Geographical Society of Australasia, established in Sydney and Melbourne in 1883, had already given help to H. O. Forbes† (1851–1932), the explorer of New Guinea.[31] This society had as its major aim the discovery and commercial advance of its own territory and the development of education in physical, commercial and political geography: there was active exploration of central Queensland and the King Country, in the northern half of New Zealand, was being penetrated. The Society issued four sets of proceedings,[32] published in Melbourne from 1883, in Adelaide and Sydney from 1885 and in Brisbane from 1886. And the Geographical Society of Quebec,[33] which issued publications from 1880, helped to open up the north with a survey of Hudson Bay, where seven winter stations were set up to report on the movements of ice. Exploration of the Labrador coast had also begun, and it was reported as indented by deep and narrow fiords, and in some places fringed by shoals extending out for several miles. On the human side, a speaker at the Royal Geographical Society[34] mentioned that Manchuria had received hundreds of thousands of Chinese from Shantung and Hopeh since 1876, who had broken up large areas for agriculture and made Mukden one of the most prosperous cities in the Chinese Empire.

Above all this was an age of imperialism. Seen at the time its dangers were hardly apparent, yet one wonders what was in Lord Aberdare's mind when he wrote this paragraph,[35] which is worth full quotation: 'To the politicians of all the great European nations the period has been one of intense interest and anxiety, connected more or less with questions of vast territorial requisitions. To the geographer the interest, although less painful, has hardly been less keen. The French in Asia and Africa—the Russians in Central Asia—the English on the Afghan frontier, on more than one border of India, on all sides of Africa, and in Oceania—and the Germans on the East and West African coasts and among the islands of the Pacific and Australasian seas—the Italians on the Red Sea—have, while pursuing measures of national policy, made large additions to our knowledge of the globe, have stimulated inquiry into others. Never—and I need hardly even except that period of emigration which precipitated and followed the break-up of the Roman Empire—has the ferment among nations been so widespread or prophetic of such great consequences. The foundations of new empires, new civilizations, are being laid over vast portions of the earth.'

Wide as the vision of the Royal Geographical Society was at this time, the new societies in Edinburgh (1884) and Manchester (1884) were mainly commercial in character, though the Scottish society early branched out into the study of physical geography, oceanography, climate, and even place-names, none of which had any obvious relevance for commercial advancement. Oddly enough, the same address[36] was given to both societies—in Manchester at a preliminary meeting on October 21, 1884, and in Edinburgh at the inaugural meeting on December 3, 1884, by Mr H. M. Stanley† (1840–1904), on 'Central Africa and the Congo basin; or the importance of the scientific side of geography'. This address, also given in Glasgow and Dundee, stressed that there should be a keen interest in geography in 'every large seaport or manufacturing town in this Kingdom, in Liverpool, Manchester,

Glasgow, Edinburgh, Newcastle, Hull, Bristol or Plymouth', from which 'ships, or products of loom or forge, are dispatched to every part of this globe possessing a mart . . . an enterprising shipowner, or an enterprising manufacturer (who) wishes to know his business well . . . should know something of geography'. Indeed, it was argued, 'geographical knowledge clears the way for commercial enterprise . . . the beginning of civilization'. Stanley noted that 'of the thirteen million square miles of Africa, only three million had European commercial links: assuming that two million square miles was valueless economically, what of the other eight million square miles'? Build railways, he advised, rather than put your faith in 'unjustifiable violence': the French, the Germans, and—in the Congo—the Belgians had shown more wisdom than the British. Stanley originally went to Africa as a journalist from the *New York Herald* but he became the major geographical lion of his time and acquired considerable fame from his book on *The Congo and the Founding of the Free State* published in 1883, as well as from his previous books, *How I Found Livingstone* (1872) and *Through the Dark Continent* (1878).

In Manchester the first expression of the need for a geographical society apparently came in 1879 at a meeting in the Chamber of Commerce to confer on the promotion of trade with Africa.[37] Dr Vaughan, the Bishop of Salford, spoke on the existence of societies of commercial *[sic]* geography in France, and other speakers urged the foundation of such a society in Manchester. A circular issued to firms and individuals produced little response, however, so the effort was abandoned. By 1884, it was said, 'a great change for the worse had come over the commercial world': meanwhile through the work of many intrepid explorers including Livingstone, Stanley and Speke† (1827–64), and also through the enterprise of the seventy continental geographical societies as advisers of governments and publishers of material, Africa was becoming known: 'the bare, barren waste of the maps of the centre of that great continent had blossomed into fertility, with waterways of thousands of miles and vast populations waiting for the birth of civilization and the olive branch of peace.' Most of the early publications of the society were articles likely to interest the commercially minded; and Stanley in his address spoke of the dangers facing the cotton trade, with Americans and several European countries manufacturing their own and India already exporting cotton fabric. Yet in the first year of its life, the Society followed the Royal Geographical in making a close study of geographical education: the response from Owens College (later part of the University of Manchester) was most encouraging, as both the Principal, Dr Greenwood and the professor of geology, Boyd Dawkins, favoured the study of geography: a course of twelve evening lectures included the 'ancient Geography of Britain' a summary of its geological development, the historical geography of England (from prehistoric times to the medieval period) and 'commerce and the colonies' (medieval trade, the age of discovery and the modern overseas expansion of Britain).

The Scottish Geographical Society was doing similar work from its foundation, only more of it. And from the beginning it attracted to its membership a wide variety of people with strong scientific interests; H. R. Mill, for example, records that he was drawn in through his interest in physical science.[38] Apart from those interested in commerce and politics, or in education, there was the appeal of the *Challenger* voyages, of which the reports, published at an office in Edinburgh, laid the foundations of modern oceanography and made substantial contributions to zoology, meteorology and other sciences.[39] The Ben Nevis observatory was opened in 1883, largely through the work of the Scottish Meteorological Society.[40] In the university there were many men of enterprise, including the brothers Archibald (1835–1924) and James Geikie (1839–1915), successively professors of Geology, and Patrick Geddes† (1854–1932) in the Botany department who was already showing his versatility and wide human interests.

Outside the university, Edinburgh already had a distinguished tradition in cartography through the firms of Bartholomew and Johnston: the Bartholomew firm produced numerous town plans, maps illustrating guide-books and geographical journals, and notably the layer-coloured maps on various scales that have served many generations of walkers and cyclists. The Johnston[41] firm sponsored a famous *Physical Atlas* (1850) which, though originally to be an English edition of Berghaus' *Physikalischer Atlas* was in fact an independent enterprise, and the *Royal Atlas* of 1859 (p. 238). Keith Johnston read a paper to the Royal Society of Edinburgh in 1851 showing the great neglect of the Ordnance Survey and constantly advocated the development of geography as part of a liberal education: he was one of the founders of the Scottish Meteorological Society. His son, A. K. Johnston† (1844–79), went in 1866 to Stanford's, London, and helped to produce the

Globe Atlas of Europe and the maps for Murray's *Handbook of Scotland*. He worked on Africa and in 1870 published the *Lake Regions of Central Africa*. Later he worked on Stanford's *Compendium*, and he spent more than a year as a geographer in Paraguay. He died during an expedition to the head of Lake Nyasa. Both Johnstons were in close touch with German geographers. At this time Edinburgh's publishing firms included A. and C. Black, then responsible for the production of the ninth edition of the *Encyclopaedia Britannica*, which included a good deal of new geographical material.

Early editions of the *Scottish Geographical Magazine* show a wide range of interest. No doubt H. M. Stanley was a member-catching magnet, but although there were many articles on remote and strange places there was much more besides. James Geikie, for example, wrote an excellent account of physical features with some condemnatory blasts at current textbooks and wall-maps. The geographical notes dealt with all parts of the world, and one on British trade with Tibet says 'at present there is practically none'. The reviews included an assessment of the last volume in Stanford's Compendium, *Europe*, by F. W. Rudler and G. G. Chisholm† (1850–1930)—the latter to do so much for geography later. Another review commented adversely on a textbook whose author said that in Scotland 'the divine side of the moral law is observed by all classes of the community'. One must mention a modest four-page article by the Rev. James Gall[42] of Edinburgh, who described his map projections, and added that he did not know if there was a copyright: in any case anyone was welcome to use them, but could his name be remembered and the projections distinguished as Gall's Stereographic, Isographic and Orthographic respectively? The Society was most anxious to assist exploration but had not the means to do so; however it issued a circular on behalf of H. O. Forbes,[43] about to go to New Guinea, which produced nearly £400. He was helped also by substantial donations from the British Association, the Royal Geographical Society and the recently formed Australasian Geographical Society (p. 57). Other societies formed in Britain included those of Tyneside, Liverpool and Southampton. The Tyneside society, founded in 1887, issued journals from 1889–1915 and again from 1936–9: the Liverpool Society was founded in 1891 and Southampton in 1897, but in both these cases the only publications were annual reports.[44]

Academic Possibilities

When the *Annales de Géographie* was founded in 1892, under the direction of MM. P. Vidal de la Blache and Marcel Dubois, some terse comments were made on the need for a new journal in a country where so many existed already.[45] Their real aim was to make a journal comparable with the *Mitteilungen* or the *Geographical Journal*, as some of the local journals, published at great financial sacrifice, gave their readers 'sous couleur de géographie, tout autre chose que de la science'. The purpose of the *Annales* was to make geography an academic discipline, to present discoveries in a form at once scientific and literary and not primarily as 'curious facts' from African or other distant lands: its readers were not to be titillated by 'nouvelles à sensation'. Academic influence was greater in the direction of the *Annales* than in the councils of the Royal Geographical Society, as the *Geographical Journal* in 1893 tactily recognized in a generous note on the work done in France.[46] The comment was made that 'The extent to which local geographical study is organized and encouraged in France contrasts strongly with the almost exclusive interest in foreign countries which sums up all British geography in the average British mind'. The 1893 note quotes a syllabus of investigation drawn up by the geographical section of the Congress of Learned Societies in France: their aims included an investigation and cataloguing of the more interesting manuscripts and maps preserved in public libraries or private collections; the delineation of the boundaries of one or more of the old provinces of 1789; the study of the lives and work of French explorers before 1789; the mapping and investigation of the actual distribution of dwellings in France, with their arrangement as individual farms, hamlets, villages and towns; the distribution of ethnic suffixes on maps; the recovery of popular names for natural features lost from the official maps; and the boundaries of old local divisions (the *pays*) such as the Brie, Beauce, Morvan and Sologne, as shown by the customs, dialects, and characteristics of the people. It was also intended to study past and current changes in the coastline and other physical changes in the landscape, and traces of the earliest people—since assiduously studied by archaeologists. Finally the term 'colonial geography' appears—surely such study was a necessity as more and more new material (some of it very raw, however) appeared each year, though this necessity was tardily recognized in Britain but not, even during the inter-war period, in Germany. Some of the work suggested in this programme has been done, but by no means all; to it we owe the French national atlas, the volumes on France in the *Géographie Universelle*, and a wide range of local geographical literature. For some fifty years, E. de Margerie† (1862–1953) was a devoted worker for the *Annales* and other geographical enterprises in France.

At the Aberdeen meeting of the British Association in 1885, H. A. Webster[47] (editor of the ninth edition of the *Encyclopaedia Britannica*) read a paper on 'what has been done for the geography of Scotland and what remains to be done'. Mr Webster spoke of the statistical accounts of Scotland, but advocated the definition of 'floral and faunal regions as they naturally exist'. Feeling his way to the regional idea, he added that there should be a survey of river basins, the length of a river, its navigability and tidal limit and—a popular idea at the time—the extent of land between chosen contours. On all rivers there should be hydrographic stations, where the observations would be related to meteorological data. A bathymetric survey of all the lakes,[48] also suggested, was achieved in later years by John Murray and others, but first tried in 1893 with Derwentwater by H. R. Mill, his wife and Edward Heawood† (1864–1949): later A. J. Herbertson† (1865–1915) helped in this work. Webster[49] also advocated the study of place-names (onomatology) in Scotland, on which a number of papers appeared: he also advocated a study of the 'rise and relations of modern Scottish counties and of the Border', a story still only partially told, and exact mapping of the burghal and municipal areas of Scotland, achieved later by the Ordnance Survey. Most interesting of all, he urged the production of distribution maps showing the density of population, birth and death rates, the distribution of trade and commerce, and the state of education; and he contrasted the fine maps prepared by Petermann in the 1851 Census with the poor ones in the *Statistical Atlas of England, Scotland and Ireland* published by Johnston in 1882. Finally, Mr Webster gave a hint to the Royal Geographical Society,[50] which, he hoped, would 'without intermitting its labours in the field of foreign exploration . . . turn its attention homeward, and see that something worthy of England is done for English geography'.

Years later, in 1896 and 1900, H. R. Mill[51] read papers to the Royal Geographical Society in which he advocated a survey of the whole country on the 1:63,360 scale, to include physical features with soils, rainfall, vegetation and agriculture, and population distribution. There was also to be an index of place-names. Mill had in mind regional memoirs, covering such areas as the Weald, the Cornwall–Devon peninsula, Wales, the Lake District, the Pennine Chain, East Yorkshire, the Southern Uplands, the Central Plain, the Highlands: these regional memoirs were to be based on short memoirs on each one-inch sheet. Quoting in support of this scheme his famous sentence 'Geography as a science is the exact and organized knowledge of the distribution of phenomena on the surface of the earth', Mill added sharply, 'the attitude of geographers has hitherto been directed mainly towards the collection of facts; we now require to discuss and arrange them.' The 1900 article on south-west Sussex, dealt with an area in which local geographical contrasts are vivid and it is perhaps not remarkable that many geographers have been attracted to the area between London and the south coast. Even so, as late as 1921 Mill[52] defended his scheme, which had been 'one of the worst disappointments in a life rich in such fiascos'. Had the scheme been a success, he maintained, by 1914 there would have been some twenty volumes, each of some 1,000 pages, covering the country: during the 1914–18 war 'the total cost would have been saved many times over in the prevention of aerodromes on impossible sites, or attempts to cultivate unsuitable land or the use of water-power to save coal'. Of course Mill's idea was not new, as reference to the Templemore memoir on p. 31 shows: in a later time much of the work he envisaged—but not all—was done by the inter-war Land Utilization Survey under geographical direction and by planners during and after the 1939–45 war.

Actual dangers lay in the geographical ignorance of countries requiring new settlers: in 1886, the Scottish Geographical Society published a communication from J. T. Wills drawing attention to the grave misapprehension fostered by government handbooks on Australia.[53] In South Australia, it was said, 'apples and pears will grow almost everywhere and oranges in many stations; the vine thrives luxuriantly, and the mulberry well. The climate generally is that of South Italy . . . suitable for the vine, olives and other fruits.' This gave a rosy view of a state with valuable agricultural resources over a strictly limited area. But perhaps the information on New South Wales was more pernicious. It was alleged that 'natural pastures exist over the whole colony but especially the western division (where) the great plains are nearly level tracts, well watered and clothed in good seasons with "luxuriant verdure"'. Some of the pastures, said to be 'unequalled in the world', carried only 150 to 180 sheep to the square mile on which, in spite of an expenditure of millions of pounds to make wells, dams and reservoirs, 'one-third of the sheep have been swept away by dry seasons in the last four or five years.' In a later article, Mr Wills[54] showed that there was, in fact, a very definite limit to settlement in Australia, and that land with a rainfall of less than ten inches was worth next to nothing. He regretted that South Australia had not published rainfall or crop returns since 1882, to conceal bad seasons: as yet, comparatively little was known of rainfall variability.

Many writers of the time stressed the continuing need for exploration and mapping. In 1893, Sir Clements Markham[55] pointed out that in spite of a vast amount of work, 'with the exception of countries in Europe, British India, the coast of the United States and a small part of its interior, the whole world is still unmapped': indeed 'the time for desultory exploration is past, and the need is for mapping'. Apart from the need for work in Southern Arabia and in Asia Minor, much of the interior of Canada was still unknown or only partially explored, and there was an obvious need for exploration in South and Central America: the Polar area was largely unknown, and Antarctica much neglected. But as early as 1887, H. J. Mackinder[56] pertinently asked 'can geography become a discipline instead of a mere body of information?' He goes on to say that the world was 'now near the end of the roll of great discoveries', except in Polar regions, on the borders of frozen regions, in New Guinea, Africa or Central Asia where 'for a time good work will be done'. Meanwhile 'as tales of adventure grow fewer and fewer, as their place is more and more taken by the details of Ordnance Surveys, even Fellows of Geographical Societies will despondently ask, "What is geography?"'. If only more of them did! Addressing Section E of the British Association eight years later,[57] he said that the British had no need for dissatisfaction with their contribution to precise survey, to hydrography, to climatology and to biogeography. 'It is rather on the synthetic and philosophical, and therefore on the educational, side of our subject that we fall so markedly below the foreign and especially the German standard.'

Mackinder was concerned to set the academic seal on British geography: he said that geographers practised three correlated arts

—observation, cartography and teaching—but in England they were good observers, poor cartographers, and teachers a shade worse than the cartographers. Fully appreciating that many vital foundations of geography had been laid in the eighteenth century (and far earlier), Mackinder[58] said that since 1870 the advance in geomorphology (the 'half-artistic, half-genetic consideration of the form of the lithosphere') had been far more rapid than 'anthropogeography', or human study. He stressed that human study could only be done efficiently in relation to geomorphology, to what he termed 'geophysiology' (oceanography* and climatology) and biogeography ('organic communities and their environments'). Mackinder urged that 'the treatment by regions is a more thorough test of the logic of the geographical argument than is the treatment by types of phenomena'. The concern was with the totality of the *environment*, a word so contentious that many later geographers have avoided it—but Mackinder never knew timidity. Communities may move from one environment to another, and even a given environment alters from one generation to another. The Normans were affected by their Norse background and the Americans built a civilization that could hardly have arisen in the Mississippi Plain. A community may remain through inertia in an area where the initial advantages have disappeared, through climatic change, by some new alignment of trade routes or for other reasons. But history is largely concerned with the migration of people who in their movement may seem to defy geographical factors—but only within limits. Many such questions came to Mackinder's mind and not the least of his gifts was his capacity to stimulate thought.

The National Geographic Society of Washington was incorporated in 1888 and published its first volume in the following year:[59] at this time, says J. K. Wright,[60] this was 'larger and better illustrated' than the Bulletin of the senior American society. The first number of the new journal included a paper by W. M. Davis on 'Geographic methods in geologic investigation': he speaks of the 'ideal' cycle of erosion, and interruptions within it, such as the waterfalls of Pennsylvania, 'points of sharply contrasted hardness in the rocks of the stream channel.' Shortly afterwards Davis published a second paper on the rivers and valleys of Pennsylvania, and several of his more famous papers appeared in the *National*

* Mackinder's *Britain and the British Seas* includes far more material on the surrounding seas than more modern geographies.

Geographical Magazine, including that on the Seine, the Meuse and the Moselle in 1896. But Davis abandoned the magazine when it became 'popular'. Also in the first volumes it was reported that the route of the Kiel Canal had been fixed, but there was to be no Channel Tunnel. Within America, the irrigation problems of California were considered; while farther south it was suggested that the whole idea of a canal across the isthmus of Panama was absurd and that the ' Compagnie Universelle de Canal Interocéanique de Panama ' would remain ' the most gigantic failure of the age '. Of the other geographical societies founded in the United States, the most notable was the Geographical Club of Philadelphia[61] (1891), which in its bulletin, published from 1893–1939, included numerous reports on Arctic exploration and articles on regional geography. Its foundation was ascribed to 'an awakening of the spirit of adventure and scientific exploration', and much interest was caused by Peary† (1856–1920), Nansen† (1861–1930) and other Polar explorers.

In 1895 the sixth international congress was held in London, and was thus hailed by the *Geographical Journal*:[62] 'an opportunity, such as has never occurred before in this country, will be presented of demonstrating the importance of geography as a science of high precision and vast extent, rich in results of theoretical interest, and of practical value, which affords an unequalled discipline in education when rightly applied.' At the opening session Sir Clements Markham spoke of the necessity of better geography teaching in the universities and schools, and after strong appeals by Mackinder and Herbertson the Congress passed a resolution on the subject. Other resolutions aimed at the encouragement of Antarctic exploration which came some twenty years later, the need for a systematic mapping of Africa, the organization of a uniform scheme of seismic observation and—most valuable to librarians and later students of historical geography—the recommendation that all maps should be precisely dated: most unfortunately this was not followed by some of the map-publishing houses. But of all the resolutions, the most significant was that dealing with the 1:1,000,000 map of the world,[63] first advocated by A. Penck (1858–1945) at Berne in 1891. Penck quoted a statement of the reigning Bartholomew that only 12 per cent of the land surface of the globe was unexplored and 56 per cent was fully surveyed: happily the scheme has had fruitful results since 1895. Far from solving great geographical problems, the congress did a few useful and practical jobs, and those who have attended such congresses will understand why.

As the nineteenth century drew to a close, who could not be stirred by the vast expansion of the known world? Of Britain and its dependencies it seemed natural to sing 'Wider still and wider shall thy bounds be set' for this was the sentiment of a great age of discovery, not unmixed with imperialism. Given a long period of peace, the world could go forward to heights of prosperity and civilization never before known. True there was a need for more careful study of the world, including mapping, regional survey and —slowly recognized—planning, to all of which in time geographers were able to make some contribution. But the real trouble in 1900 was the one so constantly noted in this chapter—the inadequacy of geographical teaching in universities and schools. Effective training in the subject could have made explorers even more competent than they in fact were; appreciation of the need for geographical knowledge could have prevented a British prime minister in 1938 from speaking of 'Czechoslovakia—a remote country of which we know little'. Had he read *Europe Centrale* by E. de Martonne† (1873–1955) he would have known a lot. The real need in 1900 was to set the academic seal on the noble work of explorers who braved the pathless wilderness. On them, as on the British Empire, the sun never set, but the great thrill of the next twenty years for many people was not the academic advance, but the explorations of R. F. Scott† (1868–1912) and E. H. Shackleton† (1874–1922) in the Antarctic.

CHAPTER FOUR

GEOGRAPHY IN THE EARLY TWENTIETH CENTURY

The physical basis; environmental determinism; the idea of the region; economic and political geography; geography in 1914

IN 1893 the Russian geographer, Kropotkin† (1842–1921) said that 'If Oxford had had fifty years ago a Ritter occupying one of its chairs and gathering round him students from all over the world . . . it would be this country, not Germany, which would now keep the lead in geographical education.'[1] It may be so: certainly the thrilled audiences at lectures on exploration suggested that people wanted to know more of the world and—still more significant for education—they wanted to understand it. In 1905, Sir Clements Markham[2] told an audience that geography could answer, partially at least, four questions: 'Where is it? What is it? How is it? When was it?' He added that a lack of geographical knowledge could cause disasters in war (who has not heard of generals who sent armies into bogs where expensive equipment sank?), losses of commerce (through ignorance of local conditions) or blunders in administration. A classic example of this last was the agreement of 1881 that the boundary between Chile and the Argentine, from latitude 52° S. northwards should pass through 'the highest summits of the Andes, which forms the water-parting'. In fact the water-parting does not follow the highest summits and the 'principal chain', mentioned in an agreement of 1893, proved to be indefinable. Happily in 1896 the principle of arbitration by Great Britain was accepted.[3]

The real need, so often stressed before, was better education in geography. Several other subjects were also seeking more recognition at the end of the nineteenth century, including English, the modern languages and the natural sciences. Had the educational reformers realized it, the hope for geography lay in the new secondary grammar schools, especially those established in Wales during the last ten years of the nineteenth century and in England rather later. But when a group of men met to discuss ways of improving geography teaching in Christ Church, Oxford on May

20, 1893, through the initiative of Mr B. B. Dickinson of Rugby, ten were public schoolmasters and the eleventh (and greatest) was H. J. Mackinder.[4] This meeting was the beginning of the Geographical Association, whose main but by no means exclusive purpose has been to develop geography teaching in schools, colleges and universities. Strongly encouraged by the Royal Geographical Society, the new association was formed to carry on the educational work of the older society.[5] Apart from the obvious wish to give geography a position of consequence in examination syllabuses, there was a clear need to decide what should be included within the subject.[6] In France the normal definition was 'the study of the Earth in its relationship to man'; and in America Guyot's 'study of the Earth as the home of man' was generally accepted, though W. M. Davis argued that geography was far more than this: 'the cleared roadway of a colony of pillaging ants becomes as properly a subject of geographical study as a railroad that connects centres of population.' And Davis underlined his point by quoting another American, C. R. Dryer† (1850–1926), who said that geography dealt with 'the distribution of every feature and the environment of every creature on the face of the Earth'. Probably no conception was wider than that of Davis, who used a term—now generally forgotten—'ontography' for 'all the responses of organic forms to their physical environment, whether in physiological structure, in individual behaviour, or in racial habits'.

The Physical Basis

In the early part of the twentieth century, as in the nineteenth, many workers were concerned with the global recognition of types of climate, with the mapping of rainfall and other data, and with the recognition of vegetational and faunal regions over the world and in small areas.[7] Polar explorers studied the limits of ice, the condition of the seas, the weather and climate, the possibilities of navigation in summer and winter and even the likely mineral resources or strategy in the event of war. Not only was there a wish to complete the exploration of the world, particularly of Antarctica, but also to delimit its major features: and global thinking became more practical as the data on its separate parts accumulated, though much generalization rests on inadequate data. One example may suffice: numerous as the local records of rainfall are in Great Britain, every year the British rainfall organization has appealed for more, expecially in remoter parts of the country. Inevitably this affects the knowledge of water resources, not only for hydro-electric schemes but for domestic and industrial purposes. But inadequacies in rainfall observations are slight in comparison with those for other climatic phenomena such as temperature and humidity, for the number of reliable stations is meagre even in Britain and a mere skeleton in many parts of the world. Neverthe-less this has not deterred many people from making maps of climatic regions, though it is understandable that their com-plexity grows as the data accumulates.

Why bother about mapping climatic data? No more convinc-ing demonstration was ever given than that of Griffith Taylor, who from 1910 was physiographer to the British Commonwealth Weather Service and showed that there were definite and ascer-tainable limits to the agricultural settlement of Australia, which apart from certain favoured areas must remain primarily a pastoral country.[8] He showed that only on the south was the inland rainfall at all reliable and commented that economic climatology is a better guide to the future of Australia than the experience of the 'man on the spot', possibly limited to a few good seasons. He showed the dangers of expanding agricultural life too far and gave warnings that might well have been noted elsewhere, notably in Inner Mongolia where Chinese settlers go forth hopefully in humid years to be defeated by the dry years that follow. And in the United States, unwise efforts to push the agricultural frontier too far west have brought disastrous soil erosion and the total removal of the topsoil from ploughed fields. From the study of climate, workers were led on the one hand to consider physical features and on the other vegetation, agriculture and human settlement.

All the systematic branches of geography were related and as early as 1893 Kropotkin[9] had said that he 'could not conceive a physiography from which man had been excluded'. On the other hand many problems of physical geography, or geomorphology, apparently had little human relevance, and W. M. Davis was clearly concerned that no area should be excluded from investiga-tion, whether trackless desert, Antarctic waste or uninhabitable sand dune,[10] not least because such areas may become of great human significance in future ages. For many decades polar ex-ploration was regarded largely as adventure but its relevance to climatology and air travel, not to mention its strategic significance, has proved to be considerable. Davis would certainly not have agreed with J. W. Gregory (1864–1932), who said[11] that geography did not 'search for underlying principles nor attempt to discover underlying causes' and was 'physiography plus a descriptive topography which cannot quite be confined to pure description' but 'must sometimes glide into explanations'. Davis was greatly concerned with explanation and it was clearly essential to his concept of the cycle of erosion in various types of landscape under different climatic conditions. His great *leit-motif* was the phrase 'structure, process and stage' but he wished physiographic descriptions to be written 'with verbs in the present tense', as the inevitable basis for human study. Davis, having been warned as a young man by the President of Harvard that he was doing no research and might lose his job, wrote some five hundred papers during his career, so direct quotation from any single work can be capped by others of apparently contrary meaning. But as Baulig[12] has said, he raised geomorphology to the state of an autonomous science by establishing its aim as 'the explanatory description of landforms', related to geology but dealing with the present rather than the past. At all times, Davis emphasized the need for selection. Writing in 1924 on the progress of geography in the United States,[13] he said that it was essential to mention 'anticlinal and synclinal zigzag ridges in a geographical account of Pennsylvania' or 'drumlins and proglacial river channels in an account of western New York' or 'the deltas of the Nile and the Ganges in accounts of Egypt and India': 'all such matters,' he adds 'enter perfectly into the explanatory treatment of the physiographic base of geography'. It is hard to imagine any satisfying account of the geography of Egypt which ignores its delta and who would fail to be intrigued by the landforms of Pennsylvania? And drumlins, once seen, are never forgotten, especially if they are divided up in fields, as in Ireland.

Generations of students have been fascinated by the concept of the 'cycle of erosion' put forward by Davis under 'normal' and glacial climates, for limestone regions, or for coasts of submergence and emergence. It is possible that Davis originally evolved his concept from a three-volume work on the geology of Wisconsin in which T. C. Chamberlin† (1843–1928) spoke of valleys in the driftless area as young or old.[14] Many workers have long since discarded the main ideas of Davis, especially in Germany, but the idea of the evolution of landscapes through cycles still appears useful though it is unlikely that there has ever been one complete cycle of erosion, as interruptions due to changes of sea-level and climatic fluctuations have been frequent. However great the criticism may be, Davis did much to make intelligible the study of landforms, or at least to provide hypotheses for later theories. His thought rested on the idea of causation or inter-relationship, and Leighly has commented that Davis's view closely followed that given by Russell Hinman[15] in 1888: 'physical geography seeks to trace the operation of the laws of nature upon the earth; upon the air, the water and the land; upon plants, animals and even upon man', or, in shorter form, 'the study of the earth in relation to man.' Fundamentally Davis's thought rested on the presumption of an unbroken chain of causation linking the physical phenomena of the earth's surface, the organic world and human society; and this was an extension of the Darwinian concept of evolution through natural selection to the intellectual and social realm.

At the beginning of the twentieth century, the outlook for geography was apparently very bright indeed, as it was ready to reap the fruits of the abundant discoveries of the nineteenth century. Many travellers had blazed trails, bringing back observations that were interesting, indeed fascinating and of vast popular appeal, yet unexplained, unclassified and uncorrelated. There was a need to pass from an empirical to a systematic approach, seen in such work was the efforts of A. J. Herbertson to establish climatic regions, of W. L. Sclater (1863–1943) to show zoogeographical regions, or of others to delineate vegetational regions.[16] In time there were numerous attempts to establish physical regions; in Africa Passarge† (1867–1958) found in 1908 that this could be done comparatively easily—he divided the continent into High Africa, Low Africa (which included the Congo basin with its surrounding highlands, the Sudan, the Sahara and the humid forest area of Guinea) and Africa Minor or the Atlas lands.[17] All these were subdivided, and as the example just given shows, partly on a climatic basis. The modern geographical pioneers who took a world view were concerned with the idea of the inter-relationship of all physical phenomena and this global view was a natural product of the times but necessarily tentative, and involving many hypotheses rather than proved facts. Much remained to be tested by detailed mapping: in 1921, for example, after researches extending over many years, H. R. Mill noted that the wettest known days did not occur in mountains, but in lowlands, and he sought an explanation in meteorology rather than in physical geography.[18]

In his work Mill established certain 'principles' (he does not call them 'laws'), notably that on the windward side of a hill the rainfall increases from the lower to the higher levels, and on the leeward side, the rainfall is greater at some point on the slope than at the summit. The correlation between mountains and areas of high rainfall, like many stimulating generalizations, has interesting and significant modifications.

H. R. Mill worked on a measurable distribution in his work on rainfall: all he wanted, like his predecessors and successors, was more data. And rainfall distribution, as for that matter any other climatic distribution, can be transferred to maps, varying in accuracy and in complexity (normally in inverse proportion) according to the adequacy of the data. Such trite phrases as 'maps are the tools of the geographer', have mentally blurred the basic fact that the vast majority of maps can only be produced by a certain amount of generalization, especially those of continents, or of the world. Nevertheless the comparison of maps showing rainfall, vegetation or physical features has long proved suggestive, nor does it end there. The present author remembers showing Kendrew's map of the distribution of olive cultivation[19] to a classical archaeologist who said, 'Why—that map corresponds almost exactly to the limits of Greek civilization.' But not of Roman influence—and there lies an interesting field of speculation.

Environmental Determinism

The idea of the unity of the world is generations old, but it became relevant to modern geography through the increasing appreciation of the interaction of physical phenomena. Preached by physical science for generations, this idea was strongly urged by Ritter,[20] but with teleological implications that were disliked by W. M. Davis and many later workers. Davis was attracted by the evolutionary idea, expressed as the adaptation of all the earth's inhabitants to the earth, rather than by the idea of the adaptation of the earth to man.[21] In a paper published in 1902, he said that geography had passed through three stages:[22] until *c.* 1800 it was 'a body of uncorrelated facts'; later it was 'teleological' rather than evolutionary and finally it was dominated by the 'causal notion', in which all phenomena occurring on the earth's surface were related. The need, therefore, was to study the inorganic environment and then 'all these responses by which the inhabitants, from the lowest to the highest, have adjusted themselves to their environment.' From this definition as from other sources, the phrase 'response to environment' has become commonplace among students: it is clearly of Darwinian origin. Davis had not the certainty of belief passed by Ritter, to whom the earth was made to be the home of mind, soul and character, in which every man had his own opportunity of serving the eternal purpose of God; and when his work was done the earth remained for those who came after, 'the advancing millions having new power to fulfil the noble purposes of human life . . . (in) a world capable of constant development.'[23] Davis was impressed by the idea of the survival of the fittest through natural selection: others looked for racial and in time psychological explanations for human actions though, as knowledge of the world grew, some looked to man as the supreme arbiter of his own destiny and others to the environment as a key to the study of human society. On the one hand, there are people dazzled by the endless hopes of human conquest of adverse and difficult environments and, on the other, those who find everything predetermined by physical factors beyond human control.

Endless argument has centred on the relation between people and environment, some of it of the 'Can you grow bananas at the North Pole?' variety. To some extent at least the success of environmental determinism as a theory was a reaction against ideas of 'man and his conquest of nature' which have been, and still are, widely held. A modern example of note is the British groundnuts scheme in East Africa which perished in miserable financial failure.[24] Less recent but none the less serious human mistakes in using land made such catastrophes as the spread of the American dust bowl through the extension of field agriculture rather than grazing, and soil erosion due to disafforestation, not only in historic areas such as the Mediterranean but also in the newer lands such as the United States or New Zealand. It may be true, as Bowman† (1878–1950) has said,[25] that no nation has ever occupied its frontier; but there is widespread evidence that there are climatic limits to profitable agricultural settlement, in spite of the production of quicker-maturing crops or hardier cattle. During the nineteenth century the problem of world food supply could apparently be solved by the steady reclamation of new lands—so marked a phenomena that the 'conquest of nature' phrase seemed relevant; in the twentieth century, the increase of food supply is achieved more and more by the application of scientific methods to areas already cultivated. But it may well be that the Soviet Union has, in fact, large areas ripe for reclamation, especially in the Asiatic territories, as well as vast possibilities of agricultural expansion by mechanization, fertilization, irrigation and other means in areas already occupied—indeed both developments are obviously proceeding with some speed.

That there are limits to human choice in the use of land no one could deny, and it is no accident that some of the statements in the above paragraph appear to confirm the determinist hypothesis. But the transformation of areas from one period to another shows that human enterprise has profound results. A mining area, sparsely occupied for many generations, may spring into active life for a few decades and then be neglected once more. The search for oil has given some areas undreamed-of wealth and strategic significance—perhaps for a time only. And the classic example of Denmark's agricultural revolution in the 1870's, after her disastrous war with Prussia in 1864, stands out as the sensible reaction to adversity of a spirited nation seizing the economic advantages caused by the growth of town populations in Britain and Germany, transforming her poorer soils by fertilization and drainage, and adopting a far-sighted organization of co-operation. Again, the skill and patience of the Dutch in adding to their agricultural area by reclamation from the sea, combined with their skilful water-engineering, stands out as an achievement of long-designed planning. And this has entered deeply into Dutch life, for drainage management necessitates the careful planning of new housing, with generally fortunate results.

The examples given above perhaps show how varied the choices of people may be from one age to another, or even at one and the same time. Yet in so far as they deal with the agricultural aspects of life in Holland and Denmark, they illustrate some comments[26] of Ratzel† (1844–1904), supposedly an arch-determinist: '"Culture" is freedom from nature not in the sense of complete release, but in that of much wider union. The farmer who gathers his corn in the barn is really as dependent on his ground as the Indian who harvests in swamps wild rice that he did not sow. We do not, on the whole, become freer from nature while we deeply exploit and study it, we only make ourselves in single cases independent of it, while we multiply the bonds . . . man . . . has profoundly changed the face of the earth.' Of course a single quotation from any author can be misleading, but Ratzel apparently saw man and the earth evolving together as a result of reciprocal influences. As farmers, fishermen or herders, men have from the beginning of human time lived in close touch with the natural environments: more and more, man has changed the face of the earth, not least by adding cities and so making, for an increasing proportion of the world's people, an artificial environment as worthy of consideration as the natural environment.

Theories of 'geographical control', or more cautiously 'influence', not uncommonly appeal to workers in other fields, just as a geographer may speak of economic or social influences a shade glibly. Indeed the determinist theory is often traced to H. T. Buckle (1821–62), a nineteenth-century historian who looked for a scientific theory of history—a pattern in fact.[27] He never heard Sir Lewis Namier eloquently explain that 'there is no pattern—only contingencies: circumstances are never the same twice'. But the economic historian, looking for an explanation of the concentration of the cotton industry in part of the Pennines, the Rossendales and valleys on their fringes, cannot ignore the work of Ogden on the distribution of abundant supplies of soft water available in numerous streams.[28] On a larger scale, many historians have seen the relevance of the geographical factors, not least George Adam Smith in his work on the Holy Land, A. E. Zimmern on the Greek Commonwealth and J. L. Myres on the Near East. Every Russian historian must say something of the vast spreading lowlands over which territorial control has oscillated forwards and backwards, of the place of the great rivers in national history and of the spread onwards into the steppes and forests of Siberia. History is not merely a matter of dynastic intrigue or diplomatic chicanery, significant as these are in its story; nor is it the puppet-like dancing of humanity to a few great geographical influencies, relevant as these prove to be.

The American writer, Ellen Churchill Semple† (1863–1932)[29] is the chief modern exponent of environmental determinism. Her work was published in the form of numerous articles and of three books, *American History and its Geographic Conditions* (1903), *Influences of Geographic Environment* (1911), which has the subtitle 'On the basis of Ratzel's system of anthropogeography', and *The Geography of the Mediterranean Region: Its Relation to Ancient History* (1931). Ratzel's work was published in 1882 and 1889 in two volumes and Miss Semple's work was more than a translation, partly because the original author regarded his work as incapable of translation into English as it 'must be adapted to the Anglo-Celtic and especially to the Anglo-American mind'. Indeed Miss Semple,[30] having begun to make an 'adapted restatement of the principles', went further and made 'a radical modification of the original plan'. Her frank preface, so generous to Ratzel, shows that she had, in fact, written a new book, and discarded many ideas that no longer seemed relevant in the early twentieth century, including, significantly enough, 'the organic theory of society and state . . . now generally abandoned by sociologists', but destined to rise again in later glorifications of the state. Miss Semple considers political science in its geographical implications: having said that 'political geography developed early as an offshoot of history', she adds firmly that 'the most fruitful political policies of nations have almost invariably had a geographical core', and, with characteristic *élan* gives as examples 'the colonial policies of Holland, England, France and Portugal, the full-trade policy of England, the militantism *[sic]* of Germany, the whole complex question of European balance of power and the Bosporus and the Monroe doctrine of the United States'. And in the same paragraph the division of England between 'the south-eastern plain and the north-western uplands' is considered with examples from the Roman conquest 'which embraced the lowlands up to about the 500-foot contour line', the Wars of the Roses, the Civil War, the Reform Bill of 1832 and the struggle for the repeal of the Corn Laws. Even political parties 'tend to follow approximately geographical lines of cleavage' and so the generalizations continue from page to page, from line to line, but with careful documentation.

It would be folly to judge Miss Semple's work solely by quotations, or to ignore her warning in the preface that she has 'purposely avoided definitions, formulas and the enunciation of hard-and-fast rules', though Nature's laws 'are none the less well founded because they do not lend themselves to mathematical finality of statement'.[31] Always regarded as the modern apostle of determinism, Miss Semple (in her own words) 'speaks of geographic factors and influences, shuns the word geographic determinant, and speaks with extreme caution of geographic control'. Would she, one wonders, have cared to be regarded as a 'stop-and-go determinist'? Her views are obviously sincere and it may be that she wrote herself into a more rigid determinism than she originally intended. Having given a warning on the dangers of quotation, it is now proposed to consider a few pages of her work. In her chapter on 'classes of geographical influences',[32] she deals first with physical characteristics, noting the Darwinian hypothesis that food supply and climate induce many slight changes in animals and plants such as size, colour or thickness of skin and hair. Still following Darwin, Miss Semple notes that people like the Tlingit Indians who live largely in caves have strong arms and chests but thin legs which, however, develop if they work in salmon canneries.[33] No single influence is operable: the people of the Auvergne, for example, a European 'misery spot', are weak, due 'in part to race' and also to a harsh climate and poor food, as well as to a 'disastrous' emigration which consists of the taller and more robust individuals. (Incidentally, it is widely accepted, though never conclusively proved, that emigration always attracts the stronger stocks.) Emigration is not a universal panacea, for the Ladak in the dry Himalayan valleys die if they move to the plains and therefore must stay at home in an area of sparse food supply: so nature applies 'the leash to drive and the leash to hold'.[34]

Environmental influence goes far beyond physical effects, and one of Miss Semple's most famous passages is on Hell, where the Jew expected to be permanently fried and the Eskimo permanently frozen.[35] To Buddha, born in the steaming Himalayan piedmont, heaven was conceived as Nirvana, the cessation of all activity and individual life. Even vocabularies were affected: for example, the Samoyeds of north Russia have a dozen words to describe the greys and browns of their reindeer. (There are twenty-four ways of saying 'to be' in Russian: is this because Russians are as introspective as their plays suggest?) Again, man is moulded by his economic and social life and, says Miss Semple, 'the effects are none the less important because they are secondary'.[36] They may, for example, be far less clear in a highly industrialized society than in the oasis settlements studied by Younghusband† (1863–1942), who met every size of social group from the 60,000 of Kashgar to the single families 'living by a mere trickle of a stream flowing down the southern slope of the Tian Shan'. On the 'great man' theory of history, Miss Semple makes some terse remarks, for 'as a rule he is a product of the same forces that made his people. He moves with them and is followed by them under a common impulse'.[37] And even their apparent eccentricities, like Peter the Great's choice of St Petersburg for a city in the middle of marshes was 'made in response to natural conditions offering access to the

Baltic nations, just as certainly as ten centuries before similar conditions and identical advantages led the early Russian merchants to build up a town at near-by Novgorod '. In short, Peter the Great gave Russia one of its fairest cities on a difficult site, but a site not more difficult than that of many of the world's great cities: above all, the choice gave Russia an outlet to the Baltic and so to the oceans of the world. Miss Semple died long before many thoughtful people wondered to what extent Hitler and the Nazi movement were the product of the German mind, and to what extent their existence was the responsibility of the whole German nation. But it is easy to ask questions that cannot be answered.

Climatic determinism is perhaps too stark a phrase for the theories put forward by Ellsworth Huntington† (1876–1947) to the effect that there have been considerable changes of climate in historic (and prehistoric) times and that these changes have profoundly influenced the history and nature of civilization. Huntington first expounded his theories in *The Pulse of Asia* (1907), which was written after prolonged Asian travels and immediately aroused wide interest. He was interested both in the long-range fluctuations and in those of short duration: in 1916, for example, he quoted S. O. Pettersson (1848–1941), the director of the Swedish Hydrographic-Biological Commission, who had noted the human effects of the period of climatic instability which culminated in the fourteenth century.[38] Evidence was drawn from the level of the Caspian Sea, the conditions of Lop Nor and other parts of Central Asia, the growth of the big trees of California and the history of the great storms, disastrous floods and unprecedently cold winters in north-west Europe. Pettersson's main discussion, however, relates to Greenland and Scandinavia; Eric the Red was little troubled by ice on his journey to Greenland in A.D. 982 and his ships sailed nearly westwards from Iceland to the coast of Greenland, southward along the coast, and thence through the strait between the mainland and Lake Farewell. By the thirteenth century it was no longer possible to take this course and in time ships did not try to reach the Greenland coast except at the southern tip or slightly northward on the west side. One of the main Norse settlements was wiped out in 1342, and another in 1418, though these disasters may not have been solely due to climatic factors. In Norway there is considerable evidence of stormy weather in the fourteenth century, with frequent crop failures and general distress (see also p. 112).

That Huntington contributed much of interest by his climatic studies in undoubted: indeed, interest in climatic fluctuations has grown steadily and some of the distinguished recent work is discussed on pp. 112–15. The widespread movements of people from the interior grasslands of Eurasia towards the margins—movements which heralded the Bronze Age—are generally believed to be due to increasing desiccation. Such changes may take place over several centuries, but the shorter-range fluctuations also attracted Huntington's attention: in 1916, he quoted with approval a work by H. L. Moore[39] on *Economic Cycles: Their Law and Cause*: 'The weather conditions represented by the rainfall in the central part of the United States, and probably in other continental areas, pass through cycles of approximately thirty-three years and eight years in duration, causing like cycles in the yield per acre of the crops; these cycles of crops constitute the natural material current which drags upon its surface the lagging, rhythmically changing values and prices with which the economist is more immediately concerned.' Various previous workers had said much the same thing, including Brunckner in 1910, in his work on the relation of rainfall to migration and economic distress. Examples could be given from many parts of the world of its relevance, not least from China with its sinister history of floods and droughts or from the United States in the 1930's.

More controversial perhaps is Huntington's work on the direct influence of weather and climate on people.[40] In a crude form, generalizations on the effect of heat and cold on people go back to early geographers of pre-Christian times. Huntington regarded the climates best suited to intellectual activity and progress as those having a well-defined seasonal pattern with frequent changes of weather and sufficient warmth and rainfall to permit productive agriculture. Excessive heat debilitates; excessive cold stupefies. In general the so-called 'temperate' latitudes are therefore likely to see the greatest advances in civilization, and notably those areas subject to the changeable weather régime imposed by frequent depressions. Huntington followed Herbert Spencer in the view that the main centres of civilization had moved into cooler climates, an idea also favoured for Europe by Ritter and, more recently, by Marion Newbigin, who noted the north-westward movement from Babylon to Athens and Rome and then to Paris and London. And more follows—such crops as wheat now have their largest yields in areas where they are not indigenous, such as the lands around the North Sea rather than the Mediterranean. But even if one can easily measure culture and civilization (and many American geographers have apparently done so with aplomb), it does not necessarily exist in its highest form where the crop yields are most bountiful and the tractors most abundant. And it is also clear that people differ so widely in their reaction to weather as individuals that generalization is difficult. But the work of Huntington opened many doors to further inquiry, not least the medical aspects of life in particular environments. A worker of a milder tone was R. de Courcy Ward† (1867–1931) who from an initial interest in detailed studies developed theories of the relation of climate and weather to human life.

The Idea of the Region

At the beginning of the twentieth century, there was in Britain no shortage of raw material for the writing of regional geography. But H. R. Mill's paper 'A fragment of the Geography of England— South-West Sussex', already discussed on p. 64 showed how this raw material could be turned into a synthetic account of one area, though Mill[41] states in his *Autobiography* that 'for one man without funds to give a true sample of a work that required a trained staff and free access to all necessary data was an impossibility'. His aim was 'to work out one typical case in the utmost detail' to illustrate his conviction that 'the key to the principles of Geography was in the relation between the solid forms of the land and the things that are free to move about over the surface'. As noted on pp. 63–4, Mill was already hoping to establish a series of regional memoirs on the geography of Britain based on an analysis of each one-inch sheet but the story of the work of the Land Utilization Survey and other enterprises with similar ideals belongs to a later chapter (pp. 167–71). Nevertheless, he had the support of a research department in the Royal Geographical Society of which he writes 'much of its work was done, or left undone, by special Committees'. But one success was clear, if now almost forgotten: in 1904, he worked with G. G. Chisholm, and H. J. Mackinder to suggest a nomenclature for the larger features of England and Wales, especially those 'without definite names for their whole extent'.[42] These were published with a map by J. G. Bartholomew and included such names as the Hampshire Downs, the East Anglian Ridge, Norfolk Edge, Northampton Uplands, the Forest of Arden, the Humberhead Levels, the Midland Gap, the Morecambe Plain, the Plain of Gwent, the Vale of Taunton and many more, of which a substantial majority are still in use. Mill had used some of these in his contribution on Britain to the *International Geography*.

Progress was slow and in some ways uncertain, but in 1905 A. J. Herbertson[43] produced his famous paper on 'The Major Natural Regions of the World' in which he remarks that 'in this country we are less tied by tradition than in some others, as there is practically no systematic geography to guide us . . . with the rise of an academic geography . . . the wider conception of geography as the science of distributions developed'. He adds that 'geography is not concerned with the distribution of one element on the earth's surface, but with all'. The part of the paper now chiefly known is the map of natural regions, later expanded in textbooks: these were based primarily on climate but with the aim of distinguishing areas by their natural vegetation which was influenced also by altitude. Herbertson used the seasonal rainfall and the critical temperatures (0°, 10° and 20° C.) and began the consideration of so many months cold (below 0° C.), cool, warm and hot respectively, familiar to many generations of students. It may be that Herbertson derived his idea of 'critical' temperatures from German examples, including one of Köppen† (1846–1940) published in 1900: he was clearly anxious to give some permanent basis for the study of natural regions, and spoke of climates 'which do not change'. Actually they change considerably, especially if one thinks in terms of millennia rather than centuries and even in detailed regional study the short-term variants may be highly significant as study of northern Scandinavia, Iceland, Greenland and Spitzbergen shows (pp. 112–14).

Herbertson's paper included maps of the critical temperatures and of structural divisions, based mainly on the work of Suess† (1831–1914); this included six groups, archaean areas, areas of other old rocks, unfolded sedimentary rocks giving tablelands or tabular lowlands, tertiary fold areas, unfolded tertiary and recent deposits. In one form or another, some such map is taught to most students and Herbertson regarded it as a valuable basis but not as itself the main tool of regionalization. In sharp reaction to many textbooks of his day, he said that 'political divisions . . . must be eliminated from any consideration of regions'; and though he recognized the relevance of human conditions he did not—at that stage—regard it as essential to the mapping of natural regions but merely as indicative of the current stage of economic development rather than of the 'possibilities of the natural environment'. He goes on, 'The density of population map is the most direct expression of the actual economic utilization of the natural region.' Before his premature death in 1915, he became more concerned with the human aspects:[44] having said that 'natural regions exist, whether man is part of them or not', he noted that city aggregates 'alter a neighbourhood and give it a new character' and also that people have 'a mental and spiritual environment, as well as a material one . . . communities restrain or stimulate initiative according to their political structure or moral codes'.

Meanwhile, in 1903, Vidal de la Blache's *Tableau de la Géographie de la France* appeared and was welcomed in a review by Herbertson as 'a contribution to the literature as well as to the geography of France'.[45] He, like many others since, found pleasure in the harmonious blending of physical and human features in the *Tableau*, and the impression of unity in the treatment of the *pays*. The book deals with the various recognizable regional units of France one by one, and shows that each has its own distinctive agriculture due to its soil and water supply and also to the economic specialization made possible by the demands of townspeople. Far from reducing the individuality of each *pays*, modern trade had accentuated it by making their agriculture distinctive. Settlement showed a clear relationship to soil and water, for in some areas it was scattered and in others in villages. Many of the *pays* had for generations been recognized as separate from but complementary to their neighbours; but not all were homogeneous, as in some there were local deposits such as limon over chalk which gave sharply contrasting soils reflected in differences of land use. The *Tableau* is a deeply human work with a firm physical base, still of value nearly sixty years after its first appearance, though revealing an agricultural France since considerably changed. From this time, French geographers published a series of regional monographs, the first of which was *Picardie* in 1905, by A Demangeon† (1872–1940), followed by Blanchard's *La Flandre* in 1906. Incidentally, the *Tableau* was part of Lavisse's *Histoire de France* and Herbertson[46] comments that 'in the great Victoria history . . . (there was) no geography at all . . .'

Early in the modern development of regional geography, there grew up two main systems of work—one based on small local areas and another based on areas covering hundreds or even thousands of square miles. And the fundamental idea was that the small area would legitimately be expected to show some distinct individuality, if not necessarily entire homogeneity, through a study of *all* its geographical features—structure, climate, soils, vegetation, agriculture, mineral and industrial resources, communications, settlement and distribution of population. All these, it has often been said, are united in the visible landscape, linked into one whole and dependent one on another. And more, every area, save those few never occupied by man, has been influenced, developed and altered by human activity, and therefore the landscape is an end-product, moulded to its present aspect by successive generations of people. The practice has therefore been to take an evolutionary view and one of the most fascinating geographical exercises (to some people) is to attempt to reconstruct the landscape as it was a hundred, or a thousand years ago. On the large scale, those who designed world regions were seeking some general key to the situation, found by Herbertson to be largely climatic, and in 1910 by J. F. Unstead and E. G. R. Taylor to be 'those outstanding differences of relief, climate and natural resources which have the most marked influence upon the development and activities of man'. An American text of wide influence on the regional basis was C. R. Dryer's *High School Geography* of 1911. In these modest but pioneering textbooks[47] the authors used a modified Herbertson scheme: could one alter an old saying and comment, 'modification is the sincerest form of flattery'? For each continent, Unstead and Taylor give a map of natural regions, noting that 'the great climatic and vegetation divisions are the chief guides and in addition the important distinction between plains and uplands or mountains must be made'. In 1905 Herbertson's regions were coldly received by the leading geographers of the day, but within a few years they were widely studied.

There has been no end to the making of 'regions', small and large; and various writers, notably Unstead[48] as early as 1916, have tried to establish 'orders' by which small regions are combined into larger ones as, for example, the various *pays* are grouped into the Paris basin, a unit comparable in size and significance with the English Lowland: both of these are regarded as part of a north European lowland. But there have also been regions distinguished on a single criterion, such as those of relief, climate, vegetation or fauna. In America some of the early Geological Survey publications have morphological boundaries, clearly marked by references to their geological foundation, and in 1896 J. W. Powell published a map of the physiographic regions of the United States which was one of the sources for the map of N. M. Fenneman† (1865–1945), first published in the *Annals of American Geographers* in 1914.[50] In the same number, W. L. G. Joerg† (1885–1952) listed twenty-one different versions of the natural regions of North America and as a twenty-second modestly added his own.[51] Efforts to make vegetational and climatic boundaries were numerous before 1900 on a large scale, and one interesting development in Britain was a series of vegetational maps[52] of mountain areas by W. G. Smith and others, and in Ireland by R. L. Praeger (1865–1953) and his friends, which classified areas by their plant associations, such as those dominated by heather, cotton grass or bilberry. In a paper on the Harrogate and Skipton district, W. G. Smith and W. M. Rankin established vegetation zones—farmland with wheat to 600–700 feet, farmland without wheat to 1,000–1,100 feet, pastures with woods of birch to 1,250 feet 'in clough and gill' and isolated trees to 1,550 feet, pine woods to 1,400 feet and a variety of moorlands culminating in an Alpine zone at 2,000–2,300 feet. In 1904, A. F. W. Schimper's (1856–1901) *Plant Geography Upon a Physiological Basis* appeared in an English translation.[53] Originally published in 1898, this book included the famous saying that the type of vegetation in the tropical and temperate zones is determined by the amount and distribution of rainfall, the humidity of the air and the movements of the atmosphere, but the flora depends mainly on heat. Both on the world scale of Schimper and others and in local areas, the study of vegetation was regarded as immensely significant—for example an anonymous writer in the *Scottish Geographical Magazine* for 1908 spoke of such study as an essential introduction to human geography, and a guide to the main lines of future agricultural and human development.[54] And in 1906, the *Geographical Journal* struck a not unfamilar imperialist note in the comment[55] of G. F. Scott Elliot (1861–1934) that European botanists had lost themselves in 'the intricacies of *Salix*, or been overwhelmed by the genius *Sphagnum*', when they could more profitably have studied the forests, grasslands or thorn scrubs of interior Africa.

More and more the idea of regional survey gained ground, and by the outbreak of the 1914–18 war, Patrick Geddes had gained a wide hearing for his views on town planning. As early as 1902 he had published a plan for a National Institute of geography and though this was never built, his views on the need for surveys and replanning were expressed in the Outlook Tower, located in the Royal Mile of Edinburgh from 1892, and in numerous exhibitions.[56] His name is inseparably associated with Victor Branford (1864–1930) who also began as a biologist but became a financier with a lifelong interest in sociology: it was his hope that the Sociological Society, founded in 1903, would carry on the traditions of the great French sociologists, Auguste Comte (1798–1857) and Frédéric Le Play (1806–82), and study the actual processes and functions of definite regional societies. In his view, anthropologists at this time were excessively concerned with primitive societies. Geddes was from 1888 to 1917 on the staff of University College, Dundee, but lectured only for one term a year: he began life as a biologist and included among his pupils Marcel Hardy and Robert Smith, whose names are associated with the mapping of types of vegetation. Early in life, especially during a period of virtual blindness, Geddes evolved theories of the interrelation of one idea to another, based on the analogy of algebraic symbols. Just as the factors of a, b and c, may be interrelated in a multitude of ways, so ideas as entities can only exist in relation to some whole. Geddes was, however, by no means a merely theoretical thinker, for his great watchword was 'survey before action' and students of his life will discover that much of his activity, conspicuously in India, was intensely practical. All will recognize that in many ways he was a man of prophetic voice, not least in his implicit warnings of the dangers of specialization in education: this was expressed particularly in his insistence on 'simultaneous thinking', by which, for example, specialism is only a means to an end and not an end in itself. Few men have had a wider field of interest and many would say that he had elements of a universal mind. His great interest, sociology, rested on geography, economics and anthropology; and all these subjects are of value only to the extent that they contribute to the one study that really matters—humanity. To Geddes such controversies as environmental determinism against possibilism seemed futile, as at one and the same time the environment worked on an organism in passive mood while the organism in insurgent mood worked on the environment. One may perhaps quote 'The human group not merely is influenced by its environment but learns to dominate it; by improving its tools and weapons it creates a store of free energy that is available for art and thought and play; and it criticizes its laws and customs and modifies its social heritage by deliberate selection. Hence human society comprehends polity, culture, and art, as well as work, folk and place'.

Geographically, the scheme of regionalization that bears the closest relation to the ideals of Geddes is that of H. J. Fleure,[57] first enunciated in 1919. Fleure pointed out that all humanity needed nutrition, or life, reproduction, new life and beyond this the good life, met by art, philosophy and other non-material pursuits. There were in the world regions of increment, such as the Mediterranean lands, where the basic needs of food and shelter were met with relative ease and time was available for other pursuits. In contrast, there were regions of difficulty, such as highlands, where natural conditions are so adverse that emigration inevitably follows, as from Scotland. But a region of difficulty may become one of industrialization by the development of mining or other resources: in a sense, such a change makes the region one of at least temporary increment. Still poorer areas include regions of hunger or privation, where existence is difficult to maintain at all, notably on the Polar fringes. In dense equatorial forests, the climate is so unfavourable to activity that Fleure grouped such areas as regions of debilitation. Other regional types include those characterized by nomadism. Probably this theory of regionalization is now rarely considered as anything more than an interesting period piece: virtually unmappable, and never adequately mapped, it nevertheless shows insight into the quality of human life from one part of the world to another. It is also a scheme with a thorough, if tacitly assumed, physical basis. Further, it takes account of the possibilities of change in any area through altered circumstances.

In such thought there is a clear advance from the simple mapping of regions by the critical temperatures of Herbertson or the structural units of the Americans, and this advance was in part due to the stimulating thought of Patrick Geddes and others. From the distribution of natural phenomena there was a natural progression to the distribution of man and his work, and the social study of man proved to be unexpectedly complicated. This perhaps shifted the emphasis from the 'natural environment' to man, and some of the generalizations of earlier workers proved to be unsatisfying. There were, during the first twenty years of the twentieth century, various attempts to make regional geography a logical and comprehensible study of the world, using data from the widespread travels of explorers. Once the era of primary world exploration was over, so that for a large part of the world there was at least a reconnaissance survey, local study would follow—as many had urged for some time. It was the local study that revealed wide contrasts within limited areas, and numerous exceptions to the generalizations so popular before 1914, and indeed afterwards. In fact there were many local studies already, not least those of the French school: it was not for nothing that Demangeon walked every lane of Picardy and acquired so intimate a knowledge of its countryside,[58] or that Vidal de la Blache visited every *département* of France. And the great Serb geographer, Cvijić† (1865–1927) (pp. 217–19) did much the same: how different from the young modern geographer who told the writer that he could do some field-work when he had his own car!

While there was much to be gained by the study and mapping of individual phenomena, such as physical features, climatic averages, crops, areas of distinctive agriculture, population densities and many more, Hettner and others were already stressing that the real aim was to consider the relations between different phenomena and their causal connections. And here lay the difficulty—here the difficulty still lies. As Hettner† (1859–1941) expressed it in 1905, 'no phenomena of the earth's surface is to be thought of for itself, but only in relation to other places on the earth':[59] this idea proved fruitful in giving geographers a world view more apparent perhaps fifty years ago than now. It is, for example, inherent in the writings of Fleure on human regions and in the stress laid on space relations by P. M. Roxby (p. 36). But, as Hettner immediately added, it is also necessary to consider 'the causal connections between the different realms of nature and their different phenomena united at one place' (Hartshorne's translation), that is, those apparent to any discriminating observer, but all too rarely discriminated, as any seasoned reader of student theses (and possibly of works by the more mature) will know. The farmer of a few score acres knows the effects of slope, drainage, soil, weather and climate on his crops and stock: he knows, too, his relation to market conditions and his use of the neighbouring town or market village.

The American geographer, C. R. Dryer,[60] argued in 1915 that 'the final cause of a natural region . . . is economic. If the industries and occupations by which men get or can get a living in a given region are not distinctive and different from those of a surrounding territory, that region, as delimited, lacks essential unity and utility'. Dryer noted that most of the existing regional schemes were based on relief and structure, or on relief and vegetation, or on climatic, zoological or ecological distributions. All these were useful, but the best guide to regionalization was economic function as it showed the influence of relief, climate and soil, which in turn affected the vegetation and agriculture. Relief affects the climate, notably in America where the Cordillera form a divide between humid and arid areas: relief also influences the streams for navigation, irrigation and water power, and has its effects on transport. But the soil is the most important structural feature. People use the opportunities of the environment according to their inherited, traditional and imported characteristics: the economy of the twentieth-century American, Dryer notes, is based as truly on grass, grain, trees, cotton, coal, iron and copper as that of the Indian was on fish, deer, flint, bark and skins.

Economic and Political Geography

Occasionally, geographical societies arrange a chamber of horrors filled with old textbooks full of statements like 'Manchester is noted for cotton goods' or 'the county town of Cambridgeshire is Cambridge on the Cam'. It would be unfortunate if students of geography ceased to know where places are and emulated the candidate who drew a map of France showing Paris, Lyons and Bordeaux but with their positions interchanged. Equally examiners find candidates with a sublime indifference to such no doubt dull details as the products of various countries or the location of their political boundaries, like the young lady who gave Switzerland a common boundary with Czechoslovakia. Part of the case for the modern advance of geography teaching in schools and universities rested on its provision of practical information, especially for students of commerce and—more theoretically, but no less usefully— economics. In Britain, no work on economic geography has had a more distinguished history than G. G. Chisholm's *Handbook of Commercial Geography*, originally published in 1889 and in its tenth edition by 1925: from 1928, it was issued in a re-written form by L. D. Stamp, and with other helpers it is still going strong.[61] Chisholm is said to have regarded this book as a millstone round his neck, as its statistics were in constant need of revision and whole new chapters had to be added to keep the work up to date. Its basis is climate and soils, with labour and transport, followed by a long section on products and on the various countries of the world. In 1908 Chisholm[62] spoke of geography as 'the branch of study which aims at estimating the value for man of terrestrial local conditions and place relations'. He quoted Keynes's view that economic laws 'involve voluntary human action' and noted that both economics and geography inevitably took into account many facts not drawn from their own studies, such as new discoveries or industrial enterprise:[63] in 1879, for example, the Gilchrist-Thomas process was introduced in the German iron industry, and in 1905 the manufacture of calcium nitrate began at Notodden in Norway by the fixation of atmospheric nitrogen. Chisholm foresaw many industrial changes, but did not forecast them.

Essentially the work of Chisholm was based on detailed regional study. He said much that matters on the subject of industrial location[64] and maintained that 'the bulkier the material in proportion to its value, the more likely it is to be industrially treated at the place of production', though not necessarily put through a number of processes: for example metallic ores were partly smelted on the spot, but afterwards sent long distances to be refined. Ores might be moved to coal, or coal to ores, and in many cases the seat of a manufacturing industry may be determined by the relative cost of fuel and raw material—for example, coalfields attract the glass and earthenware industries. Agriculturally, the remarkable feature of Chisholm's day was the opening-up of the new wheat lands, such as the Canadian prairies, which had three main positive advantages—a favourable climate, a suitable and easily worked soil, and facilities for reaching the market. A negative but vital factor was the limited opportunities of using the land for other purposes: Manitoba for example, had almost twice the wheat acreage of England, but only one-twelfth of the acreage of potatoes, less easily transported. Siberia had suitable soils, but owing to the lack of suitable communications could not be effectively used for commercial agriculture at that time: indeed its full development had to wait for the stimulus of the modern town and industrial expansion of Russia. Chisholm[65] had a singularly vast knowledge of his subject; he had courage, too, for in his introduction to the fourth edition (1903) he says 'no student of commercial geography can be unaware how many subjects there are that still await investigation, and in many cases how far the means for obtaining the desired information are lacking'. He then lists 'a few' subjects for research (in fact seventeen) many of which could still be profitably followed, including the relation between fluctuations in climate and the yield of various important commodities, the conditions of commercially successful and unsuccessful irrigation, the exhaustibility of natural advantages for any particular type of production, the gradual conversion of manufacturing industry from the lower to the higher branches, the relation of seaports to their hinterlands. Many of these are living issues today, solved only partially, if at all, by geographers, economists or economic historians. But one cannot resist the comment that he put his cards on the table, and apparently had no fear of reviewers who delight to tell authors what should have been included in their books: he gave them the list ready-made.

Political geography as taught round the turn of the century was so inconceivably arid that many writers thankfully turned away from it to regional study. Political geography need not have been so deadly; as early as 1891 Vidal de la Blache[66] managed to make his *Etats et Nations de l'Europe* interesting: he says that the idea was to show the geographical framework of the neighbouring states, none of which could be regarded as natural regions. Many years later, some most interesting works on political geography appeared, using countries as a basis, but in the early twentieth century the main stimulus came from Mackinder's 1904 paper on 'The Geographical Pivot of History',[67] published in the *Geographical Journal*: this provided, to quote Hartshorne,[68] 'a thesis of world power analysis and prognosis which, for better for worse, has become the most famous contribution of modern geography to man's view of the political world.' Mackinder's views were developed further in his *Democratic Ideals and Reality* [69] published in 1919 but ignored, it has been said, by all except a handful of people (possibly cranks) in universities, only to be republished in 1942 and carefully read by many people since then. Mackinder[70] in 1919 said that 'the three so-called new continents are in point of area merely satellites of the old continent. There is one ocean covering nine-twelfths of the globe; there is one continent—the World Island—covering two-twelfths of the globe; and there are many smaller islands, whereof North America and South America are, for effective purposes, two, which together cover the remaining one-twelfth. The term "New World" implies, now that we can see the realities and not merely historical appearances, a wrong perspective'. Within the World Island the most important strategic location was the pivot area—the northern and interior parts of the Eurasian continent where rivers flow either to the Arctic or to salt seas and lakes: with the development of railways this area could become a land power of immense significance. In 1919, Mackinder redefined the Heartland[71] to include all the Tibetan plateau and the mountainous headwaters of the rivers of south-east Asia; but he did not include the monsoon fringe (which he terms, rather oddly, coastlands), nor Arabia, nor the 'European coastlands', which included 'for purposes of strategical thinking' East Europe, with a western limit extending from Denmark to the Balkans, but excluding the Adriatic coastal fringe. To Mackinder Russia was the modern equivalent of the Mongol empire, for around the turn of the century there were fears of her expansion on the borders of Turkey, Iran, India and China. Much of Mackinder's theory has more point now than in 1914 when the Tzarist régime was becoming steadily weaker.

Mackinder argued that the West European states must inevitably be opposed to any power capable of organizing the resources of East Europe and the Heartland: he stressed the insular and peninsular character of West Europe but in 1904 and in 1919 paid little attention to the resources of America, though by 1943 he recognized that there might arise an association of powers based on the 'Midland Ocean' or the North Atlantic, with a bridgehead in France, a moated aerodrome in Britain and a reserve of trained manpower, agriculture and industries in the eastern United States and Canada. Mackinder's famous summary,[72] without which no book on the modern history of geography could be complete, runs

Who rules East Europe commands the Heartland;
Who rules the Heartland commands the World-Island:
Who rules the World-Island commands the World.

Although this was the core of his work, *Democratic Ideals* is still worth reading for its sidelights on the distribution of power at many stages of history. Initially his views were based primarily on land- and sea-power: above all he feared an alliance between Germany and Russia and advocated their separation by a group of states not unlike those established after the 1914–18 War.

Geography in 1914

Sir Edward Grey's remark that 'the lights are going out all over Europe' has often been recalled, but war can accelerate social and intellectual changes though at too great a price. If one wished to compare or, more accurately, contrast one contribution of geographers to war and peace, it would be instructive to read the British Admiralty handbooks of the First and Second World Wars. Clearly the inter-war period saw a vast expansion of geographical work, and the development of a more assured technique—and not only in Britain, for the French geographers produced their great *Géographie Universelle*, and Americans filled library shelves by furlongs rather than yards; nor were the Germans less active. But in Britain such an advance was made possible by the devoted work of the pioneers who worked in the universities and schools and established Honours courses, of which the first were at Liverpool in 1917, at Aberystwyth and London in 1918 at Cambridge and Leeds in 1919. Many British geographers were obliged to write school textbooks including Mackinder, A. J. Herbertson with his very able wife, and Marion Newbigin, the gifted Northumbrian lady who devoted her main energies to work for the Royal Scottish Geographical Society and never held a full-time university post. Even Chisholm produced a shorter version of his monumental *Handbook of Commercial Geography*.[73] These books did much to make the subject attractive and interesting in the schools and in the extramural courses various geographers, notably Mackinder, and H. R. Mill, were conspicuously successful. British geographers of the time could not live in an ivory tower of scholarly seclusion but had to go forth to the world. And they did.

As the university courses expanded, the need for more advanced texts became apparent, and inevitably there was heavy dependence on many French works, notably in some universities the human geographies of Brunhes† (1869–1930) and Vidal de la Blache, which first appeared in 1910 and 1921. Equally physical geographers leaned on the work of Suess, Penck, Passarge or W. M. Davis: a few books were translated such as Partsch's (1851–1927) *Central Europe* (1905). Of all the British works the most famous is Mackinder's *Britain and the British Seas*, first published in 1902, a work rich in general ideas and still worthy of careful study. Archibald Little's *The Far East* (1905) gathered up and systematized many observations of Sven Hedin† (1865–1952) on his travels through Asia. And in 1914 there appeared the five-volume survey of the British Empire, a compilative work edited by O. J. R. Howarth (1877–1954), Secretary to the British Association and A. J. Herbertson: this work, though somewhat lacking in cohesion, gives a fine picture of an empire never to be the same again.[74] All these books, and many more that are unnamed but not unrespected here, are still read and some indeed are already acquiring historical interest for the mark of their times upon them. The practical applications of geography in commerce were partly responsible for such works as *Economic Geography*, which first appeared in 1914, by J. Macfarlane (1873–1953). In his original preface he says,[75] 'The development of the theory of natural regions is an indication of the rapid progress which the study of Geography has made in this country within recent years. The substitution of geographical for political units has not only imparted a new interest to the subject but has given it a new value . . . but in economic geography . . . national boundaries cannot be ignored without, to some extent, losing sight of the interaction which takes place between man and his environment.' Macfarlane also notes that the economic development of a country 'is affected not only by the nature of the geographic control, but also by the political conditions which prevail.' And many similar observations have been made by later writers, nowhere more than in Europe where the political map was redrawn once more in 1919.

CHAPTER FIVE

PHYSICAL GEOGRAPHY

The growth of geomorphology; the cycle of erosion; limestone landscapes; glaciation; general comment.

IN its fullest sense this may include climatology and bio-geography as well as geomorphology which, conspicuously in Britain, has become a favoured specialization of many young geographers.

Since the days when H. R. Mill and others were advocating climatic study, climatology has been greatly stimulated by the investigation of the upper air made possible by aviation; but although the science of meteorology has made rapid strides there is still a considerable interest in general climatology, even a popular interest reflected in the widespread sale of such works as G. Manley's admirable *Climate and the British Scene*.[1] To draw a boundary between meteorology and climatology is perhaps impossible but, as shown on pp. 257-8, geographers have made a considerable contribution to climatology and in some cases, especially in wartime, to meteorology. But one of the strangest omissions among research workers in geography has been the lack of study of vegetation in areas beyond the limits of agriculture. In Britain there were some notable studies before the 1914-18 war, quoted with interest in A. G. Tansley's *The British Islands and their Vegetation*[2] (p. 86), but in more recent times the papers in bio-geography have been very few in comparison with those on climatology and geomorphology. This may be due to the marked success of plant ecology as a subsidiary aspect of botany and particularly of the admirably-written *Journal of Ecology*; but there is an element of irony in the fact that the classification of vegetation types has made little progress among geographers for many years in spite of an active pioneering stage. Geomorphology, on the other hand, has many enthusiastic practitioners in Europe and in America, though its place in geography as a whole differs on the two sides of the Atlantic.

The Growth of Geomorphology

Many distinguished European geographers began their careers as geomorphologists and later turned to the human aspects of the subject. Of these, an interesting case is J. Cvijić[3] the Serb who went to Vienna in 1899 and worked with A. Penck and the great Suess: his earlier research was on limestone landscapes in Yugo-slavia, and much of the literature still includes Serb words he introduced. Equally he has to his credit the discovery of relics of glaciation in the Balkans and work on hydrology, but he is also well known for his book, noted on p. 216 in a political context, on the Balkan peninsula which, with his own advocacy at the time, pro-vided a basis of argument for the creation of Yugoslavia: possibly the times made the man, as indeed his preface to *La Péninsule Balkanique* rather suggests. The great de Martonne,[4] until his death the doyen of French geographers, was distinguished not only for his work in physical geography but also for his general regional work, and—especially in his *Europe Centrale*—for his handling of human problems. And the American I. Bowman began as a worker on the physical side, and turned later to social and political aspects.[5] Other continental examples of geographers with physical interests that blossomed into a wider regional con-text include J. Sölch† (1883–1951) of Vienna whose work on the British Isles shows an enormous range of interest.[6]

America provided in W. M. Davis the best known of all modern geomorphologists, but also a most controversial figure, fully and frankly discussed by the Association of American Geo-graphers on the centenary of his birth.[7] Davis travelled in every continent except Antarctica, but the foundation of his work was the excellent geological survey done in the United States, where the published memoirs included a full description of the landforms. His conception of geomorphology was 'the explanatory description of landforms: geology is a science of the past (the physical history of the Earth)—geomorphology the present aspect with the main verbs in the present tense'. In a paper of 1924, Davis[8] gives an account of his method of studying 'structure, process and stage' worthy of extended quotation: 'First, the underground structure of the mass, the surface of which is the landform in question, should be announced with as much indication of the origin of the structure as may be helpful in comprehending it in its physiographic—not geologic—relations. Along with the underground structure, brief statement should be made of the surface form just before its last movement of deformation or displacement; also of the altitude with respect to the base level of erosion in which the mass was placed by that movement, with some indication of the rate and amount of the movement. Second, the nature of the processes, usually erosional and destructive but sometimes depositional and constructive, at work in modifying or sculpturing the surface of the land mass should be made clear. Third, statement should be made of the stage that these processes have reached . . . when the moved land is so far worn down by the processes then and thereafter acting upon it . . . they can produce no further change unless a new movement takes place. Finally, the dimensions of the visible forms should be stated, with special reference to inequalities of altitude, or relief, and to the spacing of valleys or texture of dissection.' Davis apparently regarded geomorphology as holding a position intermediate between geology and geography, as in his address to the American geographers,[9] also given in 1924 (at the age of seventy-four), he said that physiographers 'should separate the analytical investigation of landforms, which ought to be regarded as a phase of geology, from the non-argumentative statement of the results reached, which must be regarded as good geography'.

This statement begs many questions. In Davis's earlier career he saw in the United States, partly through his own efforts, a rapid advance in physical geography but slow progress on the human side, which came later. But when in the twentieth century human geography developed in America, so there grew up what he regarded as an absurd division between the human and physiographic aspects of the subject, for he felt that these groups of specialists needed 'a better understanding of each other's views'. As a corollary of this he argued that authors of human, economic, historical or other studies in geography should be more aware of the physical background of their work. In passing, one may note that W. M. Davis's last article, 'The Long Beach Earthquake,' drew attention to the folly of making flimsy buildings, especially for schools, in an area liable to such shocks.[10] Davis, a master of clear exposition and an advocate of sketch-drawing in the field, provides a readable and generally well-illustrated account of his findings, remarkably free from jargon. Many of his block-diagrams, and those of his disciples, are still widely reproduced and must be familiar to readers of this book: these include the much-criticized representations of high mountains before, during and after glaciation. But many later textbooks have greatly increased the number of illustrations, especially block-diagrams as a three-dimensional view is essential: the *Geomorphology* of A. K. Lobeck† (1886–1958), first published in 1939, has almost more illustration than text.[11] Davis did not, for example, give the series of block-diagrams illustrating the 'normal cycle' now widely known to students, but this was worked out from his theories by others. Many new diagrams were made by C. A. Cotton, a disciple of Davis, to whom we owe the pioneer study of New Zealand's geomorphology,[12] published in 1922, a number of articles, and finally a series of advanced texts. Cotton shared with Davis the gift of clear exposition and so, too, did D. W. Johnson† (1878–1944), whose first major work was on coasts (see also p. 102): he also provided some fine drawings. American geomorphology tests are widely known: the 1907 'Physiography' of R. D. Salisbury† (1858–1922) was widely studied for many years. A later American geographer, W. W. Atwood† (1872–1949), also wrote on geomorphology but had a wide range of interests including economic geography.

Geomorphologists have not so far achieved an independent status in the universities, for some are working in university departments of Geology and some with Geography. Their specialism, however, has attracted many workers, varying somewhat in ultimate aims. In 1950, Kirk Bryant† (1888–1950) pointed out that in Europe most geomorphologists were working in geography departments and that many of them made contributions also to the human aspects of geography,[13] perhaps more conspicuously on the continent than in Britain. In America, on the other hand, nearly all geomorphologists are housed under some geological umbrella, in survey or university departments. Having derived great profit in the earlier years of survey from field observations of a geomorphological character, some modern geologists now ignore geomorphological methods, with the unhappy result that geomorphologists are on the defensive against the geologists and over-anxious to argue that they belong to geography after all. Some indeed, say that they will fight to the last drop of blood to see that geography's link with geology does not die. One could argue that geomorphology is worth studying for its own sake and that, as several fine scholars have shown, an academic career may be as profitably spent on the study of glaciation or vulcanicity as on palaeontology on the one hand or political geography on the other.

Many people say that no geographer can work effectively unless he has an adequate training in physical geography, for the whole traditional basis of the subject is the relationship of the earth and man. The American Geographical Society's recent *Handbook on Finland*,[14] a more substantial volume than its modest preface indicates for the series of which it is a part ('neither scientific treatises intended only for the professional geographer, nor mere compendia of facts, but basic information presented with a view to creating and sustaining reader interest'), relegates the consideration of physical geography and geology to an appendix, and gives up the geographical ghost by writing first on history, then on government and thirdly on population. Early in the book, however, there are photographs of scenes that invite explanation in terms of geomorphology and one can only regret that in a generally excellent work the fundamental physique of the country—so obviously an influence on its land use—should not be made a key to the work that follows. It is in fact the very qualities of the physical environment which make Finland attractive to study, not only for its scenery but also for its modern problems such as the resettlement of refugees on farms in a hard environment. On the humblest level, physical geography can do something to show, both descriptively and genetically, what a country is like, and this is no less true of more restricted areas. Even in studies of towns, what Carlyle in a very different connection has called 'the seeing eye' for physical features may—not inevitably must— help to explain the original choice of the site, the arrangement of its roads, canals and railways, and even the qualities of its suburbs.

But this is only one part of the work of the geomorphologist. Why should one, for example, study the periglacial areas of the world, that is those areas recently subject to ice action and now free? In the first place, they show the effects of intensive frost action and strong winds, and may give valuable data on the processes of soil formation. Secondly, the colonizing activity of vegetation can be a matter of considerable interest especially as large parts of north-west Europe and North America were glaciated in Pleistocene times. But, above all, the study of areas recently freed from ice shows the erosional and depositional work of ice and water in direct, even dramatic, form. At the present time the areas of ice in the world are declining, apparently due to changes in climate, some of which are noted on pp. 112–15. As the ice-covered area of the world diminishes or increases, so there are climatic effects: in fact there is a delicate balance of scientific effects to be measured one against another. The study of glacial fluctuations gives an opportunity of closely observing the evolution of landforms within a single lifetime, both for its own intrinsic interest and in relation to climate and vegetation.

An increasing realization that climate is a fluctuating phenomenon has led to some heavy criticism of the generalizations of W. M. Davis.[15] Kirk Bryan,[16] one of the more restrained critics, noted that Davis in postulating a 'normal' cycle of erosion, under a permanently humid climate, was postulating something that never existed at all, for one cannot assume that any area has had a permanently humid climate: in fact, says Bryan, 'the interpretation of landforms rests more and more on palaeoclimatology'. Another ground of criticism rests on the lack of exact measurement, of the type found, for example, in some modern glacial studies or those on coastal districts. A. N. Strahler,[17] for example, complains that Davis's treatment was 'completely qualitative . . . I do not recall having seen a measurement of slope angle or a precisely measured slope profile in any of his publications. Neither is there any penetrating analysis of erosional processes based on mechanics of fluids or plastic materials, though his deductions seem to show an intuitive grasp of the dynamics'. Consequently, Strahler adds, geomorphology of the Davis type appears superficial and inadequate as a branch of natural science and appeals most to 'persons who have had little training in basic physical science, but who like scenery and the outdoor life'. Davis gave a great many hypotheses which later workers could use: the trouble is that very often hypotheses have been regarded as general laws. Nevertheless, much modern geomorphology rests on the idea of the cycle of erosion put forward by Davis, which advanced from infancy through youth and maturity to old age.

The Cycle of Erosion

However much the idea of the cycle of erosion is criticized, virtually all the textbooks use it in some way, if only as a teaching device: this does not make it accurate, but at least it makes it worthy of consideration. Simply expressed, it deals with the evolution of a landscape newly exposed, such as a seafloor, by the development of a river system and the natural wastage of slopes through various types of erosion. As explained on p. 74, Davis was strongly influenced by the evolutionary thought of his time and therefore the idea of growth, maturity and decay of a landscape was attractive: those who read (or re-read) his masterly presentation of this thesis in *Geographical Essays*, a collection of his papers, may be surprised to find how closely the analogy with human life is drawn.[18] In the youthful stage, for example, the rivers are changed with abundant energy, actively eroding their valleys and conveying the debris from hill-sides acquiring, with comparative rapidity, a smoother line: in maturity the processes of erosion are somewhat slower and in old age very slow indeed. The steep slopes of youth are less stable than the gentler, rounded outlines of maturity and the whole apparent effort is to smooth out all angularities into the essential peneplain (or peneplane) of old age. Perhaps Davis's choice of the word 'normal' for the cycle of erosion on homogeneous rocks under a humid climate was unfortunate: nevertheless this is partly a question of terminology as he also included the cycle in glacial and arid areas and to some extent those of shorelines in a pioneer essay, dated 1896, on the evolution of Cape Cod.[19] But the main expression of the cycle concept on coasts is the work of D. W. Johnson, whose *Shore Processes and Shoreline Development*[20] appeared in 1919; and the name of Cvijić is inseparably associated with the development of limestone landscape (see also pp. 110–11). Critics of W. M. Davis do not always adequately recognize that he wrote repeatedly of the possibilities of interruption in a cycle,[21] and gave long study to areas of folded rocks such as Pennsylvania and New England with their complicated and fascinating river systems;[22] and one of his most famous papers is on the Seine, the Meuse and the Moselle and the relations between them.

While few later geomorphologists have attempted to provide as complete a framework as W. M. Davis, many have worked on such problems as the evolution of particular river-systems: from America in 1931 came D. W. Johnson's *Stream Structure on the Atlantic Slope* which essayed nothing less than a reconstruction of the physical history of part of the Appalachians over several million years, during which a number of erosion cycles occurred. Illustrated by an admirable series of block diagrams and sections with a relatively short text, this work begins with the statement that it had long been recognized that the north Appalachians had been formerly reduced to a peneplane of remarkably low relief, then partially overlapped by an encroaching coastal plain, later uplifted with the eventual stripping away of the coastal plain cover and the mutual adjustment of river drainage to the intricate underlying structures of the strata once buried beneath coastal plain deposits. By fascinating argument and accumulated fieldwork, Johnson shows several basic principles such as the inheritance of a drainage pattern from a cover of rocks totally removed—what is generally known as superposed or superimposed drainage—and also the effect of a changing sea-level on the physical evolution of a land-scape. But perhaps the book's most remarkable feature is the discussion—the word 'discussion' is deliberately chosen in preference to 'explanation'—of the adjustment of drainage to an involved structure: this adjustment cannot be explained solely by a study of what is now seen at the surface but only in the light of geological history. At a recent conference a British geomorphologist was explaining the existence of a superposed drainage system that originally developed 600 feet or so above the present landscape: he went on to say that he had tried to convey this idea to one of those uninspired women students not unknown in universities but it had been 'above her head'—without some geological reconstructions one cannot go far in geomorphology.

As such heavy criticism has been hurled at the work of W. M. Davis, it is only fair to give some consideration to its basis and to its fruits in the hands of later workers. Above all, Davis provided a technique for those who came after, and not least notably C. A. Cotton and many other workers. And much of the early impetus came from the excellent geological reports published for various areas of the U.S.A. by the Survey. In 1924, Davis[23] noted that these reports were not only sources of information but examples of 'the rational or explanatory treatment of landforms': he mentions[24] the generalizations of J. W. Powell on the base-level of erosion and those of G. K. Gilbert† (1843–1918) on faultblock and laccolithic mountains and on general processes of land sculpture. The geysers of the Yellowstone Park and the Grand Canyon of Colorado were regarded as sensational discoveries, which fortunately were well described and illustrated. From the 1880's the use of contours by the American Survey,[25] generally with a closer interval on the 1:62,500 scale than the equivalent British maps, made the recognition of detailed landforms such as glacial features like drumlins and eskers remarkably clear; and in recent times the Survey has provided excellent explanatory accounts of the geomorphology of particular areas.

But it was not with the local detailed study that Davis was most concerned, for his purpose was rather to establish some general principles, to find a method and a terminology. The complexity of a fully-developed river system, commonly called dendritic (branching like a tree) with tributaries flowing in many directions, was given some kind of explanation by Davis and others. In 1862 J. B. Jukes (1811–69)[26] wrote what has been described as 'one of the earliest and one of the best papers on river evolution'[27] and, one may add, one of the clearest and most simply expressed. He showed that major streams were originally developed on a long-since removed surface many hundreds of feet above the present landsurface in the south of Ireland, and had maintained their courses so that they now flow through gorges 300–400 feet deep in places. These master streams flow through 'lateral valleys', which have inevitably become deeper but not necessarily broader than the longitudinal valleys, which had the weaker streams of more recent origin. The longitudinal valleys became wider, as they were in softer rocks, here Carboniferous limestones, compared with the Devonian sandstones of the gorges. The American J. W. Powell[28] in 1875 used the term 'consequent' as 'having directions dependent on corrugation', that is directly related to the form of the surface, as compared to antecedent or superimposed (superposed) which, like the north-south streams of Co. Cork may maintain themselves alike across lowlands and in gorge-reaches through east-west ridges. Powell, in his classification, did not include subsequent streams or obsequents, their tributaries, nor resequents for streams having the same line as the consequents but at a level far lower than the original surface. Davis regarded his terms as generic, rather than merely descriptive but, as H. Baulig[29] has shown, the use of generic terms implies a certainty that may be unjustified or—to quote more exactly—makes hypotheses on the origin and development of rivers which may be unverifiable or unverified. Unquestionably the physical history of many landscapes is of vast complexity, involving not one but many cycles of erosion. Nevertheless the terminology of Davis has proved useful,[30] and the term *consequent* is used in textbooks with the meaning 'related to the initial relief of a new landsurface' and *subsequent* for streams which 'starting as gullies on the sides of the primary consequent valleys, discover and explore belts of structural weakness due to softer strata, fault—or joint—planes, and shatter zones.'

Fundamental to modern geomorphology is the idea of several cycles of erosion, due to repeated sinkings of sea-level or uplifts of the land so that the same valley may show signs of several active cycles, simultaneous in development although successive in origin. Since Davis's day of active work, much evidence has accumulated of short-term changes of sea-level over merely a few thousand years: has there, Baulig asks,[31] ever been a complete cycle of erosion covering perhaps 200 million years? Davis regarded sea-level as virtually constant even though Charles MacLaren, a writer of 1842, has suggested the inevitability of changes of sea-level through the growth and decay of ice-sheets, an idea followed and developed by later writers and notably by R. A. Daly in 1910. In fact Davis never accepted the idea of several cycles. Nor does the complexity end there, for some of the most fascinating valleys of the world, such as those in high mountains, show highly varied forms: they are not at all like the smoothly gouged-out U-shaped troughs of the Davisian block-diagrams. They broaden out into wide valleys with gently-flowing streams and narrow into gorges which presumably were present before the Pleistocene ice age and during inter-glacial periods. Many hold that ice erodes most fully where it meets most resistance, so that, as E. de Martonne suggested, the result of glaciation in a valley may be the accentuation of gorges. Another great student of Alpine topography, Fritz Machatschek† (1876–1957), supported the view that in pre-glacial times the Alps were not in a stage of smoothly adjusted relief and drainage but rather of a rugged and varied character, and therefore the effect of glaciation was to accentuate the dramatic elements in their relief and drainage.[32]

Recognition of peneplanes is a crucial key to many arguments on the origin of present landforms: it is, for example, an essential part of the theories of D. W. Johnson on the Appalachians which, as noted on pp. 102–3, rest on the idea that the present landscape includes relics of past erosion surfaces produced at different altitudes during a complex physical history. And the existence of large areas at uniform heights, or in slopes tilted according to the dip of the strata, undoubtedly challenges explanation: in the Appalachians for example, both horizontal and tilted peneplanes apparently occur. Within the western part of Great Britain there appear to be peneplanes at altitudes above sea-level of c. 200 feet and 400 feet, possibly also 600 feet and 800–1,000 feet.[33] These altitudes refer to the level of the rockfloor beneath any covering of superficial deposits such as those due to glaciation, which are mentally removed, in places with some difficulty owing to the lack of data: such deposits may be more than 200 feet thick. There is some doubt about the 600 feet surface as its apparent presence may be due to the survival of ridges and hills above the 400 feet surface, probably due to the superior resistance of certain rocks to weathering. The lowlands of South Wales, for example, and those of the south of Ireland, are largely made up of surfaces at such altitudes, and so, too, is the Central Lowland of Ireland, though there the picture is complicated by a wide range of glacial deposits of varied thickness. And in the south-west peninsula of England, surfaces of 200 feet and 400 feet are strongly represented. This may seem obvious enough and—at the best—an unexplanatory statement of landforms observable fairly easily: it is a valuable key to the regional geography as with study of the glacial deposits it illuminates the conditions of drainage as they affect the soils, vegetation and farming. The most captious critic would probably admit that it is an inevitable first stage challenging explanation.

And this is the whole difficulty, vigorously stated by numerous writers such as the American J. L. Rich[34] who said that 'much of modern physiography . . . might almost be called a science without foundation'. He thought that the quest for peneplanes had been pursued without adequate study of the processes involved in the development of these surfaces. Four main lines of investigation are followed. In the first place, remnants of nearly level or gently-sloping surfaces are a valid criterion only when the possibility of a structural origin is eliminated: this presumably harks back to the initial assumption of W. M. Davis on the ideal cycle of homogeneous rocks showing virtual uniformity of resistance to weathering and also to the idea that a peneplane virtually if not completely level may be developed on an area having a rockfloor of intricate anticlines and synclines—certainly there are such cases known (for example in the Central Lowland of Ireland). Secondly, the regional truncation of strata as seen for example in the Appalachian folds or the Cincinnati arch, may result from long-continued denudation without reduction to a peneplane, if the region is uplifted either continuously or pulsatingly in such a way that uplift more than keeps pace with denudation. Here one meets another great difficulty of the Davisian scheme—can one ever assume that an initial surface is available, for during uplift the erosion processes continue and work out a system of valleys inevitably related to the arrangement of the rocks? Third, accordance of the summits of hill-tops, ridges and sloping surfaces is exaggerated and misleading when viewed in the field, on the skyline, or on profiles projected on to one vertical plane. Here one may comment that the smooth outlines of granite hills, such as those of Co. Wicklow or the Cairngorms, obviously suggest peneplanation, especially to one who knows the physical joy of walking over miles of smooth upland country; yet the essential need may be to study the method of weathering of granite and to contrast its smooth outline with the sharply-moulded outline produced by weathering on other rocks, such as schists or gneisses, which may possess greater powers of survival than the granites. Indeed, the form of the English Lake District mountains may perhaps be best explained in terms of weathering and the search for peneplanes may be futile. Finally, the accordant altitudes of isolated hill-tops, ridges, spurs and shoulders are not a reliable criterion as these features are found at the meetings of steep slopes and are therefore subject to maximum attack by the agencies of denudation and so suffer a marked loss of altitude during any considerable portion of an erosion cycle. It is also probable that these divide features stood at all times considerably above the base level to which the peneplane was being developed.

Assuming, as one must, that a cycle of erosion, or indeed any significant part of it, lasts for many thousands or even millions of years, any chosen area may be presumed to have experienced a variety of climates and therefore a variety of conditions of weathering. Much modern work in archaeology and vegetation, especially in the analysis of peat bogs, has shown how considerable climatic changes may be; and the study of recent glacial fluctuations, notably in Scandinavia, Iceland and Alaska,[35] supports the contention that crucial changes may take place within a single lifetime (pp. 111–15). And a further consideration is that dramatic spells of weather may result in rapid geomorphological action, such as a sudden change in a river course, the creation of a whole series of new gully courses in mountains, the removal of top-soil from hill-sides by sheet erosion (sheetwash), or the migration of sand-dunes inland, though the last is not infrequently a cumulative process. An article on the Cairngorm floods[36] of August 1956 shows that there were considerable movements of loose material and even large boulders on the hill-side, with marked stream erosion and deposition on the valley floors. Forces of disintegration are relatively strong in the Cairngorms, which normally have a snow-cover lasting for several months liable to carry loose material away quickly in times of melting: on the higher levels, vegetation becomes increasingly sparse and culminates in a true tundra with only intermittent plant life. Rich[37] argued that the rate of inter-stream degradation is determined by the extent and character of the vegetation and the relative resistance of the rock to weathering which, in mountains subject to frost such as those of the British Isles, may be decidedly active.

Although the weathering process may be perceptible in mountains such as the Cairngorms, at least in its dramatic phases, the arid regions of the world show the erosive processes in even more dramatic form. An obvious influence is the work of intermittent streams working out their courses in unconsolidated sand or other materials, whose powers of erosion are considerable, especially if the irregular rains are heavy.[38] There is no general base-level, such as sea-level in the 'normal' cycle, but rather a series of base-levels in depressions such as the central playa lakes or salinas (salt flats) of inland drainage basins. One widely known feature of arid areas with mountains, such as the Basin and Range province of the United States, is the wastage of the mountains by rock disintegration, of which the exact causes are somewhat controversial, to form an inclined rock floor covered in part by alluvial fans or sand and gravel debris. Pediments, associated with the decay of mountains, were first recognized by G. K. Gilbert in his work on Utah and have since been widely recognized in the deserts of North America. They may be due to intermittent but powerful water erosion, to the decay of rocks by heating and cooling, to frost action through the freezing and thawing of water in cracks in a climate liable to rapid fluctuations of temperature, and perhaps in part to sand-blast action through occasional heavy winds. It may be that several causes contribute to the erosion of the hills but the explanation is uncertain. In the arid cycle more and more debris accumulates in each basin, unless a stream manages to cut through a ridge into an adjoining, higher basin to remove part of its debris. Certainly, examination of some of the American desert sheets of the Survey shows a large number of individual, unconnected basins at widely differing altitudes, each with its own base-level.

Much of the work on arid landscapes is based on conditions in the Basin and Range province of the United States, but Davis noted also that rock deserts cover a larger area of the world than the sand and gravel deserts, interspersed by mountains, which he knew best from experience. Davis[39] defined arid areas as those where the rainfall is so small that plant growth is scanty, no basin of initial deformation is filled to overflowing and no large trunk rivers are formed so drainage does not reach the sea (in a few cases, such as the Nile, the Colorado and the Hwang, a trunk river may maintain itself across a desert). Rivers act discontinuously and weathering is largely physical—so described in a characteristic Davis passage: 'in the production of the fine waste, the splitting, flaking, and splintering of local weathering are supplemented rather by the rasping and trituration that go with transportation than by the chemical disintegration that characterizes a plant-bound soil.' Davis paid generous tribute to Passarge, the great German geographer whose work on African deserts is famed. In the African case, and probably in other old-world deserts, many of the landforms may be partly inherited from former ages of humid climate, markedly modified as aridity became greater. Yet the vast plains of African deserts, studded only with residual hills called *inselbergs*, appeared to Passarge to be smoother than any conceivable peneplane and due partly to the planing action of wind. One notable feature of the interior desert of Mongolia is the removal by outblowing winter winds of vast quantities of *loess*, an admirably fertile soil deposited in north China, as Baron von Richthofen noted during his Asiatic travels.[40] Equally the power of winds, and indeed their dangers, is known from other deserts. But one could continue to speculate for ever as geomorphologists constantly do: here the purpose has merely been to show that some apparently simple questions are, in fact, subjects for long-range research projects. No mention has been made of such problems as the development and movement of dunes or of the possible threat to cultivated lands of moving sand, not only in oases but also on the fringes of deserts such as the southern limit of the Sahara, but these and other problems have already attracted at least some research workers. Incidentally, as transport routes, tourist resorts, sources of minerals, military training grounds and for other purposes, deserts are of not inconsiderable human significance.

Limestone Landscapes

Limestones differ markedly in type and include both soft chalks and harder rocks that are self-cemented by solution into a hard, stony or crystalline condition. In England and France the chalk is a soft white limestone, with a small mixture of clay or sand and a number of flints due to silica concretions: commonly the chalk areas have a few permanent rivers and a large number of dry valleys. These are, for France, admirably shown in the drainage maps of the National Atlas. At one time it was argued that these valleys were made during the Pleistocene glacial period when the surface was frozen, at least for long periods, but as the features of the English and French chalk areas are similar, and in France such hard conditions may never have existed, other explanations have been sought. One favoured by Wooldridge and Morgan[41] centres around a progressive fall of the water-table, so that the springs were once at far higher levels than now and valleys were not at much higher levels and later abandoned. But there may be several possible explanations for the form of these valleys. Equally interesting, and of great human significance, is the presence of any capping of the chalk by beds of later date, as on the North Downs, of glacial deposits as in East Anglia or the Yorkshire Wolds, or of *limon* in the Paris basin. Such deposits provide totally different soil conditions even though in many areas they are shallow and make only a slight mark on the characteristic chalk landforms.

Far more dramatic landforms develop on compact or crystalline limestone areas, such as the karsts of Yugoslavia, the historic ground studied by Cvijić, the *causses* of France, the Pennine limestones, or those in the west of Ireland, especially in the Burren district of Co. Clare, in a wide belt of country extending north and south from Galway bay, and in numerous other areas. The main characteristic of such an area is the sparsity, even the total lack of surface drainage: in the lowland area of Ireland, mentioned above, there are very few continuous streams but a number of lakes called *turloughs* which may at times be completely dry but at others have several feet of water. There are also a number of permanent lakes oscillating in level. Represented here, too, are some watercourses which run for a few hundred yards or more in a chasm, emerging from the rock at one end and disappearing at the other: these are collapsed streamcourses. Swallow or sink holes, some of small size and 50 feet or more deep, occur, in some cases with a lake at the bottom: these grade into much larger depressions, some of which are several miles wide and generally called *poljes*. The essential feature of the drainage is that its courses bear no apparent relation to the surface forms though in places streams from subterranean channels find their way to major rivers: it is known, for example, that the Shannon is partly fed from subterranean sources. Experiments carried out during a recent drainage scheme on the Clare River, Co. Galway, showed that there was a maze of underground channels beneath the almost-waterless limestone surface and a great deal has also been learned from speleological investigations. Many of the surface forms are striking and, in areas of pronounced jointing, such as Co. Clare and the moorlands to the east of Ingleborough, Yorkshire, include fine rectangular limestone pavements. In both these areas also there are fine slabs of limestones forming craggy hills: the three Aran islands of Galway Bay consist entirely of slabs of jointed limestone. Cvijić[42] distinguished a cycle of erosion for limestone regions, in which underground drainage becomes steadily more prevalent until all the spring limestones were removed and the streams reach an underground impervious stratum.

Glaciation

The fascination of glacial study for the geographer lies in the very clear effect of past glaciations on the form of the land over much of north and north-west Europe, in and around the European Alps, and over a large part of North America. In many countries such as Norway, Sweden, Denmark and Finland, such work gives an initial key to the regional geography: no contrast of scene would be greater than the bare, ice-scraped surfaces so widely represented in Finland and the moraines and meltwater deposits prevalent in Denmark. Nor are such contrasts unique: in the west of Ireland there are many rock surfaces covered, if at all, by peat-bogs of recent origin but in the east there is an almost universal covering of moraines and meltwater deposits comparable with those of Denmark. A second reason for the compelling interest of glaciation is the possibility of seeing it as a living scientific phenomenon of the present time: the observations made over a period of some forty-odd years have shown that the glaciers of north Europe are by no means stable but fluctuating according to some meteorological causes, as yet not completely explained. It is probable that within geological time there have been three major glacial epochs, in upper pre-Cambrian, Cambrian and Permian times, each of which extended through some fifty million years and was separated one from the other by relatively mild inter-glacial periods of perhaps two hundred million years. The present, Pleistocene, glacial epoch has so far lasted for only one million years.[43]

Evidence suggests that each of the three previous glacial epochs consisted of an extended sequence of alternate glacial and inter-glacial periods, and that the climate oscillated between that of a glacial maximum, when as much as 25–30 per cent of the land area was glaciated and an interglacial minimum, when there was no permanent ice on the earth's surface. It is therefore suggested that at the present time the world climate is two-thirds of the way from a period of maximum glacial coldness to a period of maximum inter-glacial warmth: at the maximum glaciation world conditions were some 7–8° C. colder and during a period of interglacial mildness some 3–4° C. warmer than at present. But all the available evidence suggests that there are numerous climatic oscillations, some of short and some of long duration: these have attracted the attention of several workers including, in an earlier day, Huntington (p. 80) and more recently, Gordon Manley. Whatever the findings of such researches may eventually be, there can be no doubt that climatic oscillations have had profound effects on human distributions. The last glaciers of Britain probably disappeared some nine thousand years ago and a climatic optimum (first recognized in 1898 by R. L. Praeger[44] during an investigation of the fauna of estuarine clays at Belfast) occurred about 3,000–2,500 B.C. during which the winters were drier and the summers warmer than at present, and consequently trees penetrated the mountains to a height of 3,000 feet in central Scotland and the glaciers in central Norway disappeared. Degeneration of the climate followed, though there were periods of more favourable climate, notably from A.D. 400–1,000 with peaks in the seventh and tenth centuries, but there followed a decline once more to a time of marked coldness and storminess during the thirteenth and fourteenth centuries, which had adverse effects on the Norse settlers in Greenland and Iceland.

Recently, from c. 1880–1940, there has been a trend towards greater warmth and dryness of world climates, which culminated during the decade 1930–40 in a marked recession of glaciers everywhere. Some of the observations in Scandinavia are discussed below; but one must first note that while there is evidence to suggest that in Antarctica a major glacial recession was initiated not less than 10,000 years ago,[45] probably as part of a world oscillation (the glacial decline from the Pleistocene maximum), there is some evidence that at present accumulation may be greater than ice loss. Indeed, it has been suggested that climatic amelioration could lead to increased precipitation in the Antarctic, with a consequent expansion of the ice sheet. In time, when the results of the International Geophysical Year (1957) have been co-ordinated, the answer may be known: meanwhile, all that can be said is that 'there is some evidence that recent changes in atmospheric circulation have affected conditions over the Southern Ocean, but so far no observable changes in the volume of Antarctic ice have been produced'.[46] But evidence of a much more definite character comes from the Arctic: B. Halland-Hansen[47] noted that from May 1927 to May 1929 the area of ice in the Barents Sea was reduced from 430,000 sq. km. to 330,000 sq. km., and the thickness of ice in the North Polar sea diminished from an average of 365 cms. during the *Fram* voyage of 1893–6 to the 218 cms. measured by the Russian ice-breaker, the *Sedov*, in 1937–40. It is known that the southern limit of ice differs considerably from year to year; excellent maps in the National Atlas of Finland show the variations to be expected in the Baltic. Even so, the recent rise of average temperatures, especially in the winter months, though perhaps of a temporary character, has certain effects of which the clearest is the decay of the Scandinavian ice-fields and glaciers: conceivably the forests may range higher up the mountains and agricultural extension prove possible. Another effect is that cod have found their way to the waters of Greenland, and kindly offer a change of diet to the inhabitants.

Although the earliest indications of the present climatic variation go back to the last decades of the nineteenth century, the major rise in temperature in Scandinavia and Spitzbergen began about 1920: among the factors responsible are an increased northward transport of warm air from east central Europe towards Scandinavia, and the greater movement of air from the eastern North Atlantic towards Iceland and the Norwegian seas, leading to an increase in temperature and humidity in the European polar and sub-polar regions: attention has also been drawn to the increased strength of the Gulf Stream current in Norwegian and Arctic waters. Whatever the cause, the effects in Spitzbergen have been spectacular.[48] Traditionally the period of open water for navigation was only three months, but by 1930–8 it averaged 175 days, and in 1939 was 203 days, from April 29 to November 17: in 1945 the last boat left Longyearby on November 29 and in 1946 on December 5. Measurements show that the winter temperatures rose by some 8° C. between 1911–20 and 1931–8, the spring and autumn temperatures at a lower but still appreciable rate and the summer temperatures very little until the 1940's: more heat is consumed in summer for melting snow and ice.

Beyond reasonable doubt, there is now in progress a time of instability in the ice-sheets and glaciers around the north Atlantic. In Iceland and in Norway areas have been laid bare which had been covered with ice for at least six centuries, yet there is no certainty of continued recession: for example in the north-west peninsula of Iceland the Leirufjord glacier retreated rapidly until 1938, then made a sudden advance of some 1,000 m. in three years, rested stationary for four years, and afterwards retreated rapidly for two years: apparently the advance was due to the accumulation of snow during three winters of heavy fall, 1931, 1936 and 1938.[49] In Norway a series of observations from 1894 to 1912 showed a general retreat interrupted by a slight advance in 1906–7, and a further series of observations from 1933 shows a considerable retreat, though with a stationary period in 1942–3. The snow-line in the Jotunheim was c. 1,850 m. in 1919 but by 1949 was at least 200 m. higher.[50] Ahlmann has noted that various workers on the coasts of Denmark and Sweden have observed a measurable rise of sea-level, which in the 1940's was c. 1 mm. a year,[51] possibly due to glacier shrinking: there is so far no evidence that the volume of ice held up in Greenland and Antarctica, more than nine-tenths of the world's store of water in a glacial state, is diminishing, but melting of these ice-sheets could cause widespread problems. Out of long and careful study, Ahlmann[52] has said of the 'northernmost Atlantic', 'the present rapid, partly catastrophic shrinkage of the glaciers stands out as hitherto the last stage in a recession and decline which began about two hundred years ago when the local glaciers reached their maximum extension in historical time, possibly also in post-glacial time'. In Alaska a general pattern of recession has been observed since the beginning of the century, but a few interesting advances have also been recorded. In 1948 the American Geographical Society[53] began an investigation of the Juneau ice-field with a reconnaissance in the Coast mountains of south-eastern Alaska, and already some interesting observations have been published.

Gordon Manley,[54] dealing with the British Isles, has noted that the snow-beds on Ben Nevis and in the Cairngorms disappeared for the first time known in 1933, and as a result of the increased mean temperature in the summer months the 'snow-line' is probably about 1,600 m. Later in 1950, Manley amplified this statement and said that 'in regions of heavy precipitation' the snow-line would be about 5,300 feet (1,620 m.) in the Ben Nevis area, 5,900 feet (1,800 m.) in the Lake District and 6,300 feet (1,920 m.) in the Snowdon district. Presumably the snow-line would be higher in the Cairngorms than on Ben Nevis as the annual precipitation is less than half as great. Never have there been richer opportunities for the type of periglacial study noted on p. 100: even to those who, like the present author, have no intention of making a research study of glaciation but who love mountains, some knowledge of these problems can give additional interest to long days spent at high altitudes. But the glaciologist does not work only to give pleasure to mountain trampers, and many examples of the practical relevance of such work could be given. One example is chosen here. R. F. Flint[55] has described a series of investigations made by the United States Geological Survey as part of a government programme of improvement in the Missouri basin, which includes over a hundred dams across the main river and its tributaries, hundreds of miles of irrigation channels in areas only marginally arable, and the building of several power stations. Much of the work has consisted of the mapping of glacial deposits, and as a result several sheets of drift have been recognized and correlated one with another, as well as a number of end moraines. The investigations have also revealed a former drainage pattern substantially different from that of the present time, and both the past and the present drainage must be carefully considered in relation to reservoir construction.

Reference was made earlier to the case of Denmark, where solid rock appears only in two east-coast cliffs and even there is covered with drifts. In its present form, Denmark's surface shows the moulding influence of an ice-sheet decaying in late prehistoric times, and its associated meltwaters. The glacial history and present landforms are magnificently described and illustrated in the atlas[56] of Niels Nielsen and his collaborators: their studies still give the key to the agricultural geography, even though the soils have been vastly improved by fertilization and drainage and the sandy heaths have been used for forestry where reclamation for agriculture was hardly possible. Scania, the part of Sweden across the Sound from Denmark, has a landscape of thick moraines and Yoldia sea deposits similar to those around Elsinore and Copenhagen. And these same Yoldia deposits, with those of the Littorina Sea and the Ancylus Lake, have contributed much—perhaps indeed the greater part—of the fertile areas of Sweden and Finland. There are many other areas where a detailed survey of glacial and glacifluvial deposits could provide an essential key to the regional geography, especially in such studies as river-régimes, water-supplies, soils and agriculture.

General Comment

Within the compass of this chapter it has been possible to mention only a few of the subjects with which physical geography deals, and to show that it includes a great deal of controversial material. There are, however, classic studies of phenomena such as vulcanicity, notably by the great New Zealand geomorphologist, C. A. Cotton,[57] as well as in most texts on general geology or geomorphology. Equally, there have been numerous studies of coasts in various countries of which the major pioneer work was by D. W. Johnson;[58] and J. A. Steers[59] has given a thorough survey of the coasts of England and Wales. Another example is the finely illustrated review of the development of the Danish coasts in the Atlas of Denmark,[60] edited by Niels Nielsen (1949), which also shows the clear effects of the recent glacial withdrawal and the resultant adjustment of drainage on the country's landforms. A recent British work *Beaches and Coasts*, by Cuchlaine A. M. King,[61] with a fine bibliography, discusses the effects of changes of level as well as climatic influences such as winds on shorelines: much is owed to many detailed studies all over the world. But these comments raise several points of general interest.

In the first place, all geomorphology must rest on an historical foundation. The advantage of working in such an area as Denmark is that recent geological history is clearly reflected in the form of the landscape and therefore research can give a decipherable story. Part of the Danish work has been helped by the long-continued observations at the Skalling laboratory which, among other activities, has closely observed the development of sand-dunes and coastal areas in the vicinity. Basic to the work of Cambridge geomorphologists has been the observation of coastal movements on the north coast of Norfolk, at and around Scolt Head Island. There is no need to emphasize the value of such observations as those of glaciers in Scandinavia, Iceland and elsewhere.

Secondly, it is a commonplace of geomorphology that the catastrophic element may be decisively significant in the historical development of a landscape.[62] Recent examples in Britain include the floods on the east coast on January 31–February 1, 1953, when some 322 square miles were flooded, promenades destroyed, sea-walls breached and sand-dunes penetrated by the waters. Many effects of the storm were temporary and removed by repairs—in any case, not a few sea-side resorts rebuild their promenades from time to time. But some were permanent: for example at Lowestoft a sand cliff forty feet high was pushed back forty feet overnight and a cliff six feet high was pushed back ninety feet. On the Thames estuary, at Canvey island, the flooding was severe, but there the houses are built on land only fifteen feet above O.D.: a planning problem is whether houses should be built on such sites. Holland suffered severely during the same storm period. On a less spectacular scale floods such as those of Exmoor on August 15–16, 1952 can change the course of a river, initiate new gully courses on hillsides and do more geomorphological work in a few hours that might normally be done in many decades. Fortunately such occurrences generally attract the attention of geomorphologists willing to investigate the results before the inevitable repairs are carried out.

Third, the fascination of geomorphology inevitably rests on its reflection in the landscape of changes that may have taken scores of thousands of years to develop. The complaint made that W. M. Davis was virtually a crystal-gazer into the past and future may not be without foundation: but the modern preoccupation with mathematical measurement, though obviously likely to give fine results as indeed the glaciological work already shows, cannot solve all the mysteries of landforms. The statement has been made of the world that, there is 'no trace of a beginning, no prospect of an end', and for many workers, now as in the past, the interest lies in the long and visibly continuing evolutionary process. Like so much more in the development of modern geography, the real impetus came from the Darwinian scientific revolution of the nineteenth century.

CHAPTER SIX

THE REGIONAL APPROACH

Regions and regionalism; the idea of the natural region; the
problem of regional geography.

SCORES of definitions of the word 'region' exist and the
word has a range of meanings extending far beyond geography, including groups of counties or other political units
linked together for some administrative, social or commercial purpose, such as an electricity or gas supply, or even qualifying rounds
for international tennis tournaments. A newspaper with a national
sale will organize its reporting services and distribution from a
number of provincial centres; and the author was once shown a
map dividing Ireland into areas, each with its centre for the distribution of Guinness's stout, which could well have served as a
model for a newly rationalized diocesan system. In this chapter
the first section is given to a discussion of regions used for various
purposes with some consideration of the varied meanings of the
word 'regionalism', which in some hands becomes almost a form
of aesthetics; secondly, there is a section on the idea of the natural
region; and finally some consideration of the problem of regional
geography. Although the word 'regional' is inseparably associated
with geography, it is of wide general application. Much that is
written in this chapter is controversial, but it is hoped to show that
some interesting regional work has been done in various countries
with techniques adapted to local circumstances, and that much
more remains to be done.

Regions and Regionalism

Within recent years in Britain the term 'regional' has been
widely and loosely applied among industrialists and administrators,
generally for some major division of the country devised for
organizational purposes. In 1851 the Census[1] divided Britain into
a number of 'subdivisions' for statistical purposes, uniting in each
a number of counties and in 1946 these were altered into a number

of 'standard regions' used by many government ministries as a basis for their nation-wide activities.[2] To a large extent such sub-divisions or regions are used as a matter of convenience and those appropriate for the Ministry of Works, which operates everywhere, are not appropriate for the National Coal Board, whose work is restricted in area. Some services, such as Hospital Boards, have adjusted their areas to the provision of facilities in relation to population and potential needs as well as they could: others again, such as the Post Office, bear an essential relationship to railway transport services. Public services in Britain normally cover an area which is defined in terms of administrative units such as counties, county boroughs (technically equal to counties in their powers), or urban and rural districts within counties. Although a Royal Commission is dealing with administrative boundaries at the time of writing, it appears improbable that the present counties will cease to exist: indeed the efforts to alter them in 1946 were distinctly unpopular.

If a region is regarded, quite simply, as an area having some practical purpose of an administrative nature, one of the most logical arrangements ever made was the distribution of work-houses in the Poor Law Unions during the 1830's. Each workhouse was located in some town or (less usually) large village that served as the main market centre for the parishes within the Union, which might extend into two or more counties.[3] When the Rural Districts were formed in 1894 they covered much the same areas as the Unions, but with one significant difference—they never crossed county boundaries. The effect of local government legislation in England from 1888 has been to strengthen the influence of county councils in administration, in fact to turn them into utility-regions. Within the counties, the county boroughs (mainly but not invariably the largest towns) provide their own schools under an education committee with a paid secretariat. Cheshire, for example, has three county boroughs—Birkenhead, Chester and Stockport—with grammar schools for their residents; but the rest of the county is divided into ten divisions, each of which possesses several grammar schools. Until 1961, the grammar school pupil could indicate his choice and travel to a school several miles from home at the public expense, passing other grammar schools on the way. The unwisdom of this arrangement led to a change of policy, by which children are sent to the nearest school, but even now some curious arrangements remain. Only in the extreme south of

Cheshire is it easy to cross a county boundary to go to a state school: indeed the children from Stockton Heath, a Cheshire residential suburb of Warrington (a county borough in Lancashire) cannot walk across the Mersey to grammar schools but travel five miles or more each way daily to schools situated in Cheshire. Is it not strange that arrangements made for modern grammar school education should appear to be much less sensible than those made for the distressed in 1832?

In France, various schemes of regionalization were worked out before the 1914–18 war, including one by Vidal de la Blache as early as 1910 and another by the French Ministry of Commerce. Both these had the idea of replacing the ninety *départements* by about fifteen regions,[4] to be defined, it was noted in 1919, by 'topography and human and economic activities', each with a regional capital, such as Rennes, Rouen, Lille, Nancy, Dijon, Lyons, Grenoble, Marseilles, Montpellier, Toulouse, Limoges, Nantes, Bourges (or Orléans), Paris. The *départements*[5] were logically arranged divisions at the time of their creation in 1790, when Turgot said that the aim was so to define administrative divisions that everyone could reach the centre in one day. Eighty-three were formed in 1790, two added in 1793, another in 1808, three more due to the acquisition of Savoy and Nice in 1860 and the ninetieth, Belfort, in 1871. The boundaries were so drawn that the old provinces were respected when possible: for example Brittany and Normandy each made up a number of *départements*; Burgundy was divided into two, though the Yonne *département* was made of parts of Burgundy and parts of Champagne (it sounds like a cocktail *formidable*). The fifteen regions suggested for France could be regarded as appropriate to modern transport conditions, just as the *départements* were to the travelling conditions of 1790.

Germany had from its unification in 1870 a political framework that was a chaotic legacy from the past with a vast new economic and social structure for which various territorial units were made. In 1935 it was announced that all developments in settlement, commerce and industry, were to be controlled by state planning[6] and Germany was divided for this purpose into twenty-three planning regions, of which three were the Ruhr, Berlin and Hamburg: in fact they were defined mainly by grouping together political units in such a way that they formed economic units.

In Britain, the work of C. B. Fawcett† (1883–1952) in 1919, *The Provinces of England*,[7] attracted considerable attention, not so much perhaps for the details of the actual rearrangement he suggested, but for the general idea of a new regional division of England, indeed of the whole of Britain. Fawcett laid particular stress on the role of the major provincial city, much as Vidal de la Blache and others had done for France, and the idea of the regional capital became steadily entrenched. In France, Paris was at all times supreme; so, too, was London in England as the Census Commissioners of 1851 recognized.[8] But they regarded the county towns as centres where the heads of the chief families could congregate periodically; they also spoke of the towns used for weekly marketing visits spaced at intervals of nine to fifteen miles one from another, virtually all of which were centres of Poor Law Unions. When railways covered Britain and France, the larger regional capitals became steadily more significant; it is on such great provincial towns that the regional schemes of Vidal de la Blache, of Fawcett and of Nazi Germany were based. Some form of regionalization is regarded as a matter of social necessity, but it will be noted that all the schemes are compromises as they give some recognition to previously-existing administrative units.

The term 'regionalism' has almost more overtones than the term 'region': it is, for example, quite widely used of novelists whose work is placed in particular areas, such as Thomas Hardy in Dorset (Wessex), the Brontë sisters in west Yorkshire, the earlier Howard Spring in Manchester, George Eliot in the Midlands, and others. But this is only the expression of the local individuality that today several thinkers wish to see developed: in Britain during the 1914–18 war and for some years afterwards the *Making of the Future* series of books by Sir Patrick Geddes[9] and his sympathizers brought forward the idea of regionalism as a way to avoid uniformity of culture: Geddes, for example, urged that the real curse of Germany had been the declining influence of such gentler Germans as the Bavarians and the steady Prussianization of its life and notably of its universities, so widely venerated before the 1914–18 war. More was needed of the liberal spirit apparent in France. In an age when, inevitably, more people were living in cities than ever before, Geddes saw that the first loyalty must be local and then regional, ultimately national and, one might hope, international, and that the real evil of the time was the uncritical worship of the State; far from imperialist in his mind, he gave many young geographers of his day (and many other people) an outlook that was at once world-wide and parochial. So much of this is commonly accepted by the citizens of a later age that it is hard to appreciate the thrill its earlier presentation gave, or the significance of its stimulus to sociology. Nevertheless the essential problem, to develop in people some regional loyalty, remains though the detailed study of social adjustments has become more and more a subject with its own expertise. Indeed, two American writers, Odum and Moore,[10] in 1938 spoke of 'a considerable and growing body of knowledge about the region gathered through tested methods of research and study' by geographers, anthropologists, ecologists, economists, historians, political scientists and sociologists! Quite simply, differences between one part of the world and another, one part of a country and another, are the concern of a wide range of students and it is therefore surprising that at least one geographer should have spoken of regional geography as 'putting lines that do not exist around areas that do not matter', or that others go so far as to say that 'there is no such thing as areal differentiation'. Apart from the academic hyprocrisy of regarding any area as of no consequence, any discriminating traveller through a country must observe that there are great differences in character between one landscape and another.

Recognition of regions has become a matter of social utility, as Odum and Moore demonstrated in the United States. They quote R. D. McKenzie's definition[11] of a region as 'a geographical unit in which the economic and social activities of the population are integrated around a focal economic and administrative centre'. This definition epitomizes the idea of town and country relationship and interdependence which was inherent in the drawing of the Poor Law areas of Britain, in the suggested schemes of Vidal de la Blache in France, and of C. B. Fawcett in England and of the German planning districts during the Nazi régime. The inherent idea was to define a region for trading purposes both from its core and its circumference with due regard for travelling conditions: in recent years, motor-bus services have been used to delimit similar 'regions', both in Britain by F. H. W. Green and others,[12] and in Sweden by workers at Lund University.[13] These workers do not necessarily use the contentious word 'regions' of the areas they regard as tributary to central towns, but take refuge in such phrases as 'urban fields', 'market areas', 'hinterlands' though their ideas have much in common with those of R. D. McKenzie. Not everywhere is the motor-bus service time-table a reliable guide, however: it would clearly mean little in the United States with its high percentage of cars to households, or in the republic of Ireland with its lack of local bus services. Nor is the idea of the regional capital applicable in all circumstances: an argument recently advanced for the multiplication of universities in Britain is that each has its natural 'catchment area' of students and therefore one should be established in each regional capital, but in fact the universities of Britain have become national institutions (in varying degrees) drawing their students from homes distant as well as near. Much more could be made of the argument that a university may provide intellectual and social services to the region in which it is situated.[14]

In the United States, Odum and Moore have noted five general types of regions established for various social purposes.[15] Of these the first is the natural [*sic*] region such as a mountain range, a river valley, or a lowland (or parts of these) such as the Ohio River valley, the Mississippi valley, the Tennessee valley, the Dust Bowl and the Great Plains: in the first three an obvious need is control of water for the prevention of soil erosion and the checking of floods, the latter both for rural and urban communities. In the Dust Bowl and the Great Plains planning problems have arisen through agricultural practices of varying wisdom in relation to the physical features: as is well known, the Dust Bowl is a conspicuous case of nature hitting back. Odum and Moore used the term 'natural' as synonymous with 'physical', but many geographers have spoken of natural regions in terms of both physical and human features. Secondly, there is the metropolitan region,[16] very much a product of the present century, such as New York and its environs, the St. Louis 'region', Washington and Baltimore, and others: in Britain such areas are known as *conurbations* and have been in active growth at least since 1800 and in some cases even longer. At present, as in the early 1920's, there is a cult of the 'city region', that is the area tributary to, and closely influenced by, the great city. In any case, most geographers have agreed with Odum and Moore in regarding the world's great urban areas as themselves regional entities. Thirdly there are regions not easily amenable to precise definition, but possessing local loyalties and cultural traits: virtually this is literary and aesthetic regionalism, which gives us novels 'racy of the soil' to quote Odum and Moore[17] who say also that 'Regionalism in literary production consists in presenting the human spirit in every aspect in correlation with its immediate environment'. Equally there is an attractive field of inquiry available on the distribution of architectural types such as Celtic stone-working mingled with Romanesque architecture in the west of the British Isles or, as continental European geographers have shown, house and barn building types. The fourth group[18] consists of regions for convenience of administration, for which in U.S.A. (to quote) 'we have "regionalized" our nation and subregionalized and districted our states, our counties and our cities'. Over a hundred bureaux, departments and other agencies of the federal government have sets of regions of varying shapes and sizes as an aid to more efficient administration, or for other reasons sometimes not so apparent. Equally a wide range of regional units, apparently of *ad hoc* origin in many cases, is used by ecclesiastical and social organizations, athletic interests or mail order companies. Last of all in the United States, there are groups of states which may cover in varying degrees most of the other types, such as the New England Planning Board or the North-west Planning Board.

Regions obviously exist in abundance and proliferating complexity: in England, for example, such areas as the Midlands are recognized to exist by everyone. But they are hard to define, though some would begin by including the major towns such as Birmingham, Leicester, Nottingham and Derby, and even Oxford, though Cambridge is generally regarded as belonging to East Anglia. Yet in the 1946 classification, the term Midland is used (without the qualifying adjective 'western') for Shropshire, Herefordshire, Staffordshire, Warwickshire and Worcestershire; Oxfordshire becomes part of a 'Southern region', and the North Midlands, though including the historic counties of Danish origin —Leicester, Nottingham and Derby—with Northampton, also strides eastward to the coast and includes Lincolnshire. To the permanent chagrin of the geographers of Nottingham, there is no standard East Midlands region.[19] And the standard regions are only one of a large number of schemes that exist for various purposes.

In the United States, in 1938, Odum and Moore drew attention to the confusion in official regional boundaries: 'The traditional regional units of the country suffer dismemberment at the hands of federal regional planners more often than not. The case of New England, perhaps the most traditional regional unit in the nation and nicely set off by geographical factors, is illustrative. Of ninety-three schemes in use, only twenty are composed of the six states east of the Hudson river. Eleven others divide New England into two or more regions. In forty-one cases the New England States are grouped with larger units, sometimes with ten north-eastern states, sometimes with even larger groupings. Eight schemes cut across the New England boundary.' As noted on pp. 85–6, the United States early attracted workers who distinguished its major physical units: inherent in the argument of Odum and Moore is the recognition that so-called 'natural regions' of the physiographic type were by no means homogeneous socially. And no physical region showed this more clearly than 'the Appalachian Mountain Region',[20] which 'comprehends, in New England, in New York, and in Pennsylvania, many of the highest indices of civilization and wealth at the same time that the lower reaches include some of the most isolated and limited folk of the nation, to which factors are added great contrasts in climate, great distances in travel, great distinctions in culture and history, such that the test of homogeneity suggests the folly of making this an administrative and planning unit for human, cultural, political and economic ends . . .'

It is not essential that every 'region', defined for any purpose, should be homogeneous, nor is it intended to convey the impression that because the Appalachians are a physical unit they are a suitable unit for all other forms of study. But physical differences,[21] notably those of climate, affect commerce directly: for example cars have climatic adjustments, such as the heater of the north, the canvas water-bag of the south-western desert, the special carburettor and gear adjustments of mountain districts or the air filters of sandy coasts. Economically one area differs from another widely in income per household,[22] with inevitable effects on the amount of money spent on services and entertainment, and the market research organizations of the States consider a wide range of geographical as well as economic factors in defining their regions for sales purposes.[23] Odum and Moore list fourteen points generally considered: significantly the first of these is topographical conditions, with railways and motor routes; others of a geographical vintage include the normal locations of warehouses, the trading areas of wholesale distributors, the area a man can work from his home daily or by returning only at week-ends; also included are indices such as income tax returns, car registrations and magazine circulations, and the directly economic inquiries such as the previous volume of sales, the number of customers, the estimate of spendable money and the volume of business needed to be profitable. In such deliberate economic regionalization much is at stake and America has perhaps gone further in the systematization of such material than Britain.

However regions for any human purpose are conceived, their definition invariably involves some compromise with political boundaries. Obviously the states of U.S.A. are unsuitable geographical units in themselves, yet they are enshrined in countless regional schemes such as the New England states, the Southern or the Mountain states. In more detailed work it is possible to use smaller units, such as counties, parishes or the equivalent in various lands; but any regional exercise which involves statistical mapping must use material made available for defined administrative units. And here the distributional aspect can give life to statistics and show, for example, on an atlas sheet the effective distribution of population. In 1906 Sten de Geert† (1886-1933), the Swedish geographer,[24] began to experiment with mapping the distribution of people where they actually lived, no longer content with the assumption (still not abandoned by some) that 1,000 people in a parish will be evenly spread over its twenty square miles. Of course they may be, but in rural Sweden it is not improbable that they will be spread through a relatively restricted area of cleared farmland and that all the rest of the parish will be forested. In short, generalization can only be profitable if it is based on the local scene: neither statistics nor distribution maps in themselves are more than useful tools. Too often regional study has been discredited by its superficiality and its readiness to generalize on inadequate premises; the macrogeographical picture must rest on the microgeographical basis. Both in the physical and in the human aspects of regional study detailed local survey is essential to establish any general principles, if such there can be. And a task of even greater difficulty is the relationship of apparently different sets of physical and human factors.

The Idea of the Natural Region

Here the emphasis is on the word 'natural' as a basis of discussion, assuming that readers will accept the idea that some form of regionalization is useful for administrative and economic purposes and that the discerned individuality of particular areas may give a special vigour to literature including drama, art and even music. At times geographers have spoken as though the recognition of the natural region was their main contribution to learning and, as shown on pp. 82-90, the initial correlations in the nineteenth century between sets of physical factors, such as climate and vegetation or soils and agriculture, have contributed substantially to an understanding of the world. Mediterranean life, with its traditional agricultural emphasis on corn, wine and oil, has evolved in association with a particular type of climate, having its chief rains in the winter or, towards its margins, in spring and autumn, though the areas of similar climate in the new world have a commercialized agriculture developed mainly within the past century. The monsoonal climates of Asia give rains adequate for the nourishment of great rivers and innumerable tributaries diverted for many centuries to irrigate rice-fields in the river valleys and widely-extending basins, which in China and India have become the most densely-peopled agricultural areas of the world. Physical features, climate, vegetation and land use show a general correlation that is interesting and by now widely known. Yet the concentration of dense populations on fertile irrigated lowlands depends entirely on the maintenance of water-control, as many students realized when they read Marion Newbigin's *Mediterranean Lands*, published in 1924, with its descriptions of Egypt and Mesopotamia at various stages in history.[25]

As an idea, the natural region in some form is widely accepted. Many maps so labelled prove to be, as in America, of areas distinguished on a physical basis—for example, on p. 123 it will be noted that Odum and Moore, not writing as geographers, conceived 'natural regions' as the major physical region of America. In Britain, A. J. Herbertson, like Vladimir P. Köppen, based his scheme on climatic factors, but with a clear recognition that these were related to the distribution of physical features: modifications and refinements of these schemes are numerous in atlases and textbooks. In 1936, Derwent Whittlesey† (1890–1956) noted that Herbertson's natural regions bear an embarrassing resemblance to a map of climatic regions and he suggested a scheme of agricultural regions based on a statistical and quantitative rather than on an empirical and qualitative assessment.[26] Work of similar aims had been done by various writers in the journal *Economic Geography*,[27] founded in 1925, and also in the *Geography of the World's Agriculture* by V. C. Finch† (1883–1959) and O. E. Baker† (1883–1950), published in 1917 at Washington, which deals mainly with commodities. The Swedish geographer, O. Jonasson published a lengthy and careful study of Europe's agricultural regions in 1925–6, and in 1926–35 O. E. Baker produced a comparable study of North America which dealt with the major agricultural belts, winter wheat, hay and dairying, cotton, truck farming. C. F. Jones dealt with South America, Griffith Taylor with Australia, S. Van Valkenburg with Asia and H. L. Shantz† (1876–1958) with Africa: Shantz also worked on the soils and vegetation of Africa with C. F. Marbut† (1863–1935). Many maps have appeared which show the distribution of particular crops—in fact they are included almost invariably in school atlases; but Whittlesey and others, notably Hartshorne and Dicken[28] in 1935, had a wider aim —to show the general economic and human complexion of farming practice. Whittlesey's scheme[29] of 1936 recognizes the obvious influence of climate and weather, but shows also that agricultural life depends on the use of crops and livestock, alone or in combination, on the methods of cultivation or stock-rearing involved, on the use of land, capital and organization, or the disposal of produce, and not least on many social structures connected with farming in general. A primary classification of agriculture gives four major types over the world: first, animal-rearing dominant in areas too cold, rugged or remote for successful crop raising; second, crops dominant, animals of minor significance or even absent, with in some areas multiple cropping due to the lack of any check to growth by cold (this is a key factor in much of China and India); third, crops dominant, animals minor, but the crops grown limited by the natural environment (including soils) or by the market; fourth, roughly equal production of crops and animals, perhaps best seen in the mixed farming of middle latitudes.

Whittlesey's fourfold classification is a masterly effort of synthesis, but he goes on immediately and rightly to say that it is far too 'coarse-meshed', as many other factors must be considered before a world regional scheme of agriculture can be made. Of these the most significant is the relative abundance of land, labour and capital. Seven acres in Belgium or Holland may be a carefully cultivated market garden, using a heavy capital outlay and much labour: seven acres of equally rich land in eastern Europe may be a subsistence holding feeding the farm family but with little surplus produce for sale. If capital is not available, knowledge of scientific techniques limited or non-existent, tools rudimentary and markets undeveloped or too remote through poor communications, then the economic standard is inevitably depressed. At least from the days of Chisholm, economic geographers have drawn attention to the marked contrasts in yield per acre of crops between eastern and western Europe, that is between developed and under-developed lands (in modern terminology). Dudley Stamp[30] in *Our Undeveloped World* has gone further and drawn attention to the low yields of the machine-run, labour-scarce farming of parts of Canada and the United States. No doubt enough has been said to make it apparent that the simple 'natural region' based on climate and physical features, correlated with vegetation (often called 'natural' though modified almost everywhere by human influence), is crude, or at least elementary. Much scorn has been poured by writers on the maps of the United States which show a cotton belt, a spring wheat, a hay and dairying, a winter wheat belt and others; but, as R. B. Hall[31] has justly commented, 'the northern limit of the American Cotton Belt varies according to the price of cotton and with new techniques of production, but the Cotton Belt remains'.

This view, however, is contested by Merle Prunty[32] who in 1951 showed that the area planted to cotton in the south-east of the United States has declined almost by half, to become of secondary importance in much of 'what was the old Cotton South'. On the other hand the production per acre had increased in the cotton regions of the Georgia-Carolina Piedmont and coastal areas and the Tennessee and Mississippi valleys, so that there has been 'an eastward shift in the median centre of production, which is now situated east of the Mississippi river'. Having pointed out that 'there are seven cotton regions in the south-east today', Prunty concludes that 'there no longer is a "Cotton Belt" in the sense that the term generally has been used, and the realities of the current distribution of cotton culture indicate that the term should be discarded'. One result of the Land Utilization Survey of Britain, carried out in the 1930's (see pp. 168–9) was to confirm many generalizations long held, such as the existence of belts of arable farming in East Anglia and Lincolnshire and in North Cheshire with South Lancashire, or of areas predominantly of grassland farming in the remainder of Cheshire and areas to the south in Shropshire and Staffordshire with the Welsh borderland. In spite of the rehabilitation of British farming since this survey was made, the essential specializations still survive though the farming practices have been changed by the application of new scientific methods, more fertilization and the greater use of machinery with a smaller labour force.

Some kind of regionalization based on land use is clearly

significant, since in terms of area agriculture is the most widespread occupation of the inhabited world. And so far as one can judge in the age of the first moon-rockets, it is still the most essential occupation. For these two crucial reasons, if for no others, one must view with interest any such world scheme of agricultural regionalization as Whittlesey's[33] which, necessarily generalized, covered the globe in thirteen main types. Of these the first is nomadic herding in areas too dry to produce crops, and the second livestock ranching, notably with the cattle, sheep, goats and horses introduced by Europeans to Australia and New Zealand. The third, shifting cultivation, is mainly found in the tropical rain forests and involves migration but a low density of population (estimated by many workers as 1-2 per square mile). More closely settled areas give a fourth type, with fixed villages but the use of changing plots, described as 'rudimentary sedentary'. A fifth type is the intensive tillage, chiefly and even entirely for subsistence, with rice dominant in south and east Asia; and a sixth type is similarly intensive tillage in areas such as north China, but without paddy rice due to a shorter growing season or less rain: with this sixth type Whittlesey links the oasis cultivation of Egypt and of deserts. The seventh type, commercial plantation tillage, is essentially European in conception, and found for tea, sugar, rubber, bananas and other crops: it grades into the specialized horticulture of Hawaii, Jamaica and other districts of the West Indies and some South American oases, for example in Peru and the Argentine.

Mediterranean agriculture provides an eighth and historic type. The ninth type, commercial grain farming, best developed on natural prairies, is essentially a 'creature of the Industrial Revolution' with much machinery, a sparse population and—until recently—little fertilization: areas used for this type of farming might conceivably be transformed into the practice of the tenth type, commercial livestock and crop farming, characteristically western European, with grain production and the sale of animal products as in Denmark and parts of Holland: this grades into two other types, one poorer and one richer. The poorer type is the subsistence crop and stock farming of eastern Europe, with low yields and primitive methods now—so far as one can judge—being greatly changed; the richer type is specialized dairying, especially near cities and also in a number of other areas such as parts of Holland and the British Isles, where the stock are fed largely on grass with bought feedstuffs: in Denmark the main production is of dairy staples but the land is used mainly to grow the feed for cattle who spend seven months a year indoors. The Irish dairying industry, however, relies mainly on pasture, as the stock can be kept outside for at least seven months a year and in the milder districts almost the whole year. Finally, the thirteenth type is specialized agriculture, seen in many forms including the vineyards of Europe, the vegetable gardens of Brittany and south Cornwall, the glasshouses and allotments of the Netherlands, the rich market gardens of the Côte d'Azur and the Rhone valley, the orchards, hop gardens and soft fruit areas of England and many more. In North America horticulture is found in a wide variety of natural environments, on the sandy soils of the Atlantic coastal plain, the Mississippi valley and the Gulf Coast, the Rio Grande valley, with irrigated areas and oases west of 100° W. including those of the lower Colorado valley and of the Mediterranean climatic affinities, especially in California. This last type of farming covers a relatively small area of the world but is clearly of vast economic significance: not confined to any one climatic type, it includes areas carefully tilled for many centuries such as some rich enclaves of the Mediterranean and some continental European vineyards, but it also includes some areas to which irrigation has only recently been applied under the stimulus of an expanding urban market.

Europe and North America were divided into 'agriculural regions' by R. Hartshorne and S. N. Dicken[34] in 1935 'on a uniform statistical basis'. They note that both the U.S. Department of Agriculture and various private bodies had worked on 'types of farms' and cite the detailed study of farms based on the 1930 census, which mapped and classified the whole into 812 districts. The farm schedule of 1930 had three questions—the value of crops, livestock or animal produce sold or traded, the consumption of farm products by the household and the receipts, if any, from tourists and boarders, These returns showed types of farming varying from cash grain to self-sufficient subsistence but the 800-odd districts proved amenable to grouping into a small number of 'agricultural regions', and the maps of such regions proved generally similar to those published by O. E. Baker[35] in *Economic Geography* from 1926. Hartshorne and Dicken make the point that the agriculture of Canada and most of the United States (not the south) has a strong resemblance to that of Europe in the use of land for crops of pasture, the use of crops for foodstuffs and feedstuffs and even in the methods of cultivation. They divide the agricultural landscape of both continents into eight main types, and give each a boundary on a statistical basis: here they meet the difficulty that figures may be available only for counties or provinces too large for effective use. They group the Po valley of Italy as belonging to the maize—wheat—livestock belt (the stock being dairy cattle and pigs) though other crops include chestnuts, grapes and even rice: it is generally agreed that the Po valley cannot be classed as Mediterranean. Far different, however, is the flank of the Alps immediately north of this basin around the Italian lakes for there, as Kendrew[36] commented in his *Climates of the Continents*, 'lemon groves and olive trees flourish, signs of a Mediterranean climate.' But this area could not be statistically separated owing to the limitations of the administrative boundaries.

Mediterranean agriculture both in its traditional form of cultivating corn, wine and oil crops and in its modern commercialization for citrus fruits, raisins, figs, dates and early vegetables, is devoted primarily to the growth of human food crops. Hartshorne and Dicken mention that wheat is the main field crop, and they find an outer limit in the area having at least 15 per cent of all cultivated land (excluding hay) under olives, citrus fruits, nut orchards and vineyards—generally in the true Mediterranean areas of France, Spain and Italy far more of the land is so used, from 25 to 60 per cent. The second type—corn, wheat and livestock farming—has maize as the main crop with wheat, oats and barley as subsidiary crops; in U.S.A. this region has at least 20 per cent of its land under maize, and at least 30 per cent under maize and wheat. It is separated from the specialized tobacco and cotton areas by having less than one-fifth of the arable area in tobacco and cotton on less than half the acreage of maize. As noted on p. 129, cotton-growing fluctuates considerably according to price movements. In this type of farming the sales may be highly varied; for example in eastern Illinois commercial grain farming predominates with three-quarters of the land sown, but in the level plain of Iowa there is a great concentration on beef and pigs and here, as in parts of the British Isles, lean cattle are bought from poorer areas for fattening. This type of farming may be highly commercialized but it may also grade off into a subsistence type, for example in the southern Appalachians, the Ozarks, the Carpathians and the Balkans.

Outside the climatic limits of maize the third type, small-grains and livestock farming, has more land in crops, generally wheat and rye, with oats, barley and potatoes than in pasture and meadow. The crops are variously used for foodstuffs and as feed-stuffs for livestock supplying both dairy products and meat. But there are local variants—for example some of the *pays* of the Paris basin have a heavy concentration on wheat, oats and vegetables for the metropolitan markets (see p. 134). A fourth type, hay-pasture farming, has the purpose of producing dairy and beef cattle, pigs, sheep and poultry: the area in pasture and meadow is far greater than that under tilled crops. Much of Ireland and the western part of Britain comes in this category, and in Scandinavia it is found with some use of forest pastures. Very different is the fifth type, extensive commercial grain-farming, in which one or two grain crops cover vast areas of chernozems or comparable soils, for example in the western interior of North America, the Ukraine, the north Caucasus and western Siberia. Wheat for sale may cover as much as three-quarters of the land, and minor crops include oats and barley chiefly for stockfeeding, but also for foodstuffs and for sale. Other types include truck farming and commercial orchards, extensively developed in America and in small but important areas of Europe, the commercial-grazing areas of America in areas of low rainfall, and some 'quasi-plantation' areas of the New World, notably for such crops as tobacco and cotton.

These classifications of regions are of interest as they are indicative of landscape variations and population distribution; coming from American workers they appear to support the dictum of C. R. Dryer[37] in 1915 that 'the final cause of a natural region . . . is economic' (p. 89). But there is the difficulty that broad classifications are at once revealing and concealing, for within any agricultural region there may be a multitude of variants due to local changes in soil, variations of physical aspect or drainage conditions and the like. It seems to the present author that in all efforts to give neat presentations of natural regions far too much stress has been laid on the need for homogeneity. This may, in fact, persist for hundreds of miles in such areas as the Canadian prairies or the Australian desert, through which one can travel by train for a complete day with no clear change of scene: it may also be characteristic of the *taiga* belts of the world which are thoroughly dominated by forest having agricultural clearings as a subsidiary if vitally important economic and social characteristic of the landscape. But there are many landscapes, especially perhaps in western Europe, that are of mixed character: in Britain, for example, many of the coal-mining areas are by no means the industrial areas of some people's imagination but exist rather as a number of mining villages and—less usually—towns scattered round a landscape that is predominantly agricultural. And even in many agricultural landscapes having no mining or industry, heterogeneous elements exist in association: much of the central Irish lowland is a physical mixture of ground moraine and other glacial features used for farms, generally small in size with fields of a very few acres, interspersed with peat bogs on the one-time fens of a more favourable climatic epoch than the present day. To the extent that this landscape has any unity, it lies in the association of these two elements—the glacial drifts available and used for farming and the peat bogs used for fuel. Yet as a whole such an area is quite distinct from the virtually driftless limestone lowlands in the west of the Central Lowland, or from the glacially scoured, erratic-strewn wastes of Connemara, where virtually all the fields have been made by collecting and manufacturing soil with seaweed, sand and fertilizers.

If one can recognize areas distinct one from another in landscape, due to structure, relief, qualities of drainage, soils, vegetation such as forest, heath or scrub, agriculture, population distribution, why not say so? It can hardly be accidental that regional geography received a great impetus in the Paris basin, where Vidal de la Blache[38] and his students recognized a varied series of *pays* of which some were given to sheep-rearing, others to stock-raising or dairying, others again to crop-growing; in them the people had found the best use of their land to be that most suited to the soil and climate. This seems obvious enough, but it involves a specialization made possible by the demands of Paris and by the exchange of goods between one *pays* and another: it also involves a long-standing economic advance from the subsistence land of farming to one based on trade, or at least on the exchange of surplus commodities. Not all the *pays* are uniform in quality, for some have woods on poor sandy soils unsuited to farming, and others have *cavités douces*, richly supplied with streams and springs, in areas of limestone or chalky sheep or cattle pastures. Many, too, have *côtes* or gently scarped edges which carry vineyards near their northern European limit. Various authors have applied techniques similar to those of Vidal de la Blache in England and have found, for example, that in the south-east there is at least some degree of correlation between physical features and agriculture: C. C. Fagg and G. E. Hutchings,[39] showed that in the south-east each geological province, such as the gault, greensand and Weald clay, had its own characteristic agriculture and—looking backwards—its own natural vegetation. Although the North Downs and the South Downs are comparable in surface form, the former are partly covered with superficial deposits used for grass and woodland or even heath, and the latter are sheep pastures, but not cattle as 'the pasture is not rich enough' though 'the soft spring turf makes chalk grassland a favourite training ground for race-horses and hunters'.[40] Crops can be grown with careful fertilization. For East Anglia,[41] P. M. Roxby made a regional scheme on a basis of soil conditions and drainage and argued that the main effects of the agrarian and industrial restrictions from the eighteenth century had been to accentuate the correlation between natural conditions and farming practice, though from about 1875 conditions had been unfavourable to the East Anglian farmer due to the import of relatively cheap foreign grain and meat, except during the 1914–18 war.

Study of such regional units was regarded by P. M. Roxby as a definite contribution to future agricultural planning, for as early as 1913 he wrote with admiration of William Marshall (1745–1818) who gave most of his life to agriculture chiefly—and characteristically for his time—in estate management.[42] In his *Rural Economy of the West of England*, Marshall said that '*natural* not *fortuitous* lines are requisite to be traced; *agricultural*, not *political distinctions* are to be regarded'. He also observed that 'A *natural district* is marked by a uniformity of soil and surface, whether by such uniformity a marsh, a vale, an extent of upland, a range of chalky heights, or a stretch of barren mountain be produced' and an *agricultural* district shows uniformity or similarity of practice—grazing, sheep-farming, arable management or mixed cultivation or some particular product—dairy-produce, 'fruit-liquor': the last was apparently cider apples in Herefordshire. Marshall divided the Severn valley into the Vale of Berkeley and the Vale of Gloucester with the Vale of Evesham as its extension, and describes for each the surface, the special climatic features and the nature of the soil and subsoil—the soil 'a deep rich loam, fitted by intrinsic quality for the production of every vegetable suited to its specific nature and the latitude it lies in'. Marshall recognized that county boundaries are not divides, for the dairy district based on north

135

Wiltshire extended into parts of Gloucestershire and Berkshire and the eastern margins of Somerset. Roxby argued that rural England should be divided into natural regions on a basis—be it noted—of relief, geological formation and climate, and that for each such natural region there should be a study of the agricultural evolution since the Agrarian Revolution and of the existing agriculture and population trends. But the Roxby 1913 article belongs to an age of geographical optimism, in which the definition of 'natural regions' seemed relatively easy. Much that Roxby hoped to see was done later by the Land Utilization Survey of Britain, the modern Domesday Book, but the reports were arranged on a county basis for the greater part of Britain, partly to enable the available statistics to be effectively used. In spite of wartime difficulties, however, the various county reports, the 1:63,360 maps, and the generalized maps on a scale of 1:625,000 have proved to be a notable contribution to regional geography.

As a student of regional geography, Roxby[43] saw two things clearly. First, he followed the French regional geographers in believing firmly that no area could be understood except in an historical context: it was not for nothing that many of the first university geographers of France and Britain were trained as historians. The natural region was an end-product, moulded by many generations of people, changing the landscape from one age to another, building, tearing down and raising up houses, taking away the forests and reclaiming heaths, dividing common lands into fields, introducing new crops and new rotations of crops, breeding new stock and even making two blades of grass grow where one grew before. In 1925, Roxby[44] said that 'A physical unit tends to become an economic unit, and the more developed the means of communication, the more pronounced its regional specialization'. Yet in some areas, the demands of a great market may stimulate a type of production different from that most appropriate to the local conditions: for example in south Essex dairy-farming paid even though feedstuffs were imported and the natural conditions would appear to favour crop-farming. Working on this theme, Roxby stressed that 'the factor of intrinsic conditions and the factor of space-relationship sometimes "pull" in different directions and create complex conditions'. In all his work Roxby urged that no regional unit should be thought of as existing in isolation but in relation to a wider area: full allowance was made for changing economic conditions, not least those due to the effects of new transport facilities. American geographers not infrequently speak of 'natural landscapes which become cultural landscapes', using the word 'cultural' to mean influenced by human action; but in western Europe the human imprint is so deep that one can only reconstruct the natural landscape with difficulty. Second, Roxby[45] regarded it as a duty of the regional geographer to comment on possible future uses of land as, for example, he did in writings on China or L. D. Stamp does in *The Land of Britain*.[45] Not all geographers share this view: some might agree with an American geographer[46] who said (in 1936, but it could be said now) that 'in view of the admitted immaturity of our technique and methods, and in view also of the complexity associated with reasoning in terms of the composite, it probably is fortunate that we have not encouraged prediction'. This comment implies that the first essential task is to develop geographical work and then see how it may be applied.

From a country very different from Britain or U.S.A.—Finland —there has come a most interesting regionalization based on a four-fold classification: landforms, water, vegetation and settlements. This was originally used in the 1928 Atlas of Finland, and has been recently discussed in the Geographical Society of Finland's *Suomi*,[47] a general handbook on the geography of Finland published in 1952. There was little need of modification, though one category in the vegetation group (C—spruce and broadleafed tree swamp) was left out in later revisions of the map: the 1952 book notes that 'vegetation is the most important *equalizing* factor' in Finland, as forests interspersed with swamps and bogs are universal except in the far north, and even there the true tundra is interrupted by woods grading from conifers to birches and finally field birches of bush size. The forests differ markedly in detail and have been classified according to their ground vegetation and the age of the trees. The general classification is arranged thus:

LANDFORMS: I, high mountainous country—variation in height of at least 200 metres; mountains at least 200 m. in height predominate. II, mountainous country—height variations under 200 m., predominant heights 50–200 m. III, hilly country—height variations under 50 m., hills 20–50 m. predominate. IV, hillock country—height variations at most 20 m., predominant hills 10–20 m. V, flat lowland country—gradual slopes with little variation in contours, rises less than 10 m. VI, plain. VII, valley and table land—level, sharply delineated valleys, separated by long gradual slopes or table-top hills with relatively smooth crowns bounded by terrace formations with steep slopes.

WATER: 1, level water. 2, flowing water with rapids. 3, lakes. 4, lakes in chain formation. 5, coast and archipelago waters.

VEGETATION: A, forest. B, fjeld birch forest. C, spruce and broadleaf tree swamp (no longer used). D, pine swamps and treeless bogs. E, cultivated and meadow land. F, rocks and treeless fjelds.

HUMAN: a, 'strung-out' habitations. b, clustered habitations. c, scattered habitations. d, very sparse habitations. (Note that 'c' may include very small villages.)

For each of these categories a regional map was made, which divided Finland (with its present boundaries) into forty sections based on landforms, thirty on water, forty-nine on vegetation and twenty-nine on settlement, which has a clear relation to agriculture, forestry and rural service facilities in villages. The boundaries for all these four categories were brought together to give the 'geographical regions'—to quote J. G. Granö,[48] 'the results of these analytic maps are brought together on a synthetic map'. Only the sea provided a perfect boundary, and the regions were found to be 'weakly outlined as individual units' but 'often homogeneous as geographic complexes of a specific type'. Thus an area distinguished on a landform basis might prove to be like its neighbour in vegetation, and settlement. Nevertheless it was found possible to divide the country into sixty-five tracts, which could be grouped into sixteen territories (1952—the 1928 atlas had 104 tracts and nineteen territories). For each of the regions, the density of habitations per square kilometre was worked out, and all the available statistical material was used and—when possible—carefully mapped. Granö, having noted that the statistics were given under communes, notes that 'The boundaries of the communes deviate in some places quite considerably from the boundaries of the natural regions, but as a whole the communal statistics are also useful in regional geography.' For each tract, a landscape formula is given, for example:

1. The Turku (Abo) settled coastal area.

    III 5 AEF bc— hills 20–50 m., coast and archipelago waters, forests, cultivated and meadow land, some patches of treeless field, clustered and scattered habitations.

2. The Suomenselkä hillock area.

    IV 23 DAF ca—hills of 10–20 m., flowing water with rapids and lakes, pine swamps and treeless bogs, forests, rocky treeless fjeld, scattered and 'strung-out' habitations—this last meaning isolated farms at intervals along roads.

3. The Farther Lapland birch country.

IV II 24 BF d—hillock country with hills of 10–20 m., but also some mountains at least 200 m. high, flowing water with rapids and lakes in chain formation, fjeld birch forest and tree-less fjelds, very sparse population.

These three areas have been chosen from south, central and north Finland and it will be seen that for each the landscape formula gives an excellent shorthand impression of the area—especially to anyone who has visited Finland. Granö[49] shows that it is possible to divide Finland into two parts—settled or agricultural, and unsettled, divided by an imaginary line from the north-east shore of Lake Ladoga to the north-east of Lakes Pielinen and Oulujarvi to the head of the Gulf of Bothnia: naturally there are undeveloped areas in settled, or peninsular, Finland, and developed areas in the rest, but 'all centres of population of any importance, with the sole exception of Rovaniemi, and all railways except two lines, are within the boundaries of peninsular or agricultural Finland as above delineated'. Peninsular Finland 'is, with the exception of certain barren peat lands, a settled area . . . in the sense that small patches of fields and meadows and pasture lands and winding grey roads everywhere break the monotony of the ubiquitous dark-green conifer forest and leave signs of human effort on the scene. In districts close to the sea, particularly in the south-west and in south Ostrobothnia, where fertile areas are of considerable extent' and where the population density is comparatively high, 'cultivated expanses with picturesque clusters of buildings and winding roads completely dominate hundreds or even thousands of square miles. Here, accordingly, the natural landscape has been transformed into one with a distinctly agricultural appearance'. Finland's problems of finding living-space for one-tenth of its people from the territories absorbed by Russia have made these regional studies valuable, and they are also a possible basis for the planning regions it is proposed to establish. Arising from the study of settlement, one writer hopes that more villages and small towns will be established in the country-side.[50] The geographical work found not only in this regionalization, but in a vast amount of other work has become an effective contribution to national development. Nevertheless, it is doubtful whether any landscape formula as simple as that for Finland could be established elsewhere, least of all in a country possessing such scenic variety as Great Britain. Nor is there in Finland any vast industrial area or any large city except Helsinki; and the agriculture has not the variety found in such a country as France or Italy.

The method of regional classification may well differ from one country to another with profit. The Russian geographers have given thought to the great soil belts of their vast plains, each bearing some relation to the type of vegetation, ranging from tundra through forests to steppe and desert, each belt having agricultural possibilities and limitations. But as vast new industrial areas with large industrial settlements are developed, or new irrigation projects fertilize hundreds of square miles, the initial regionalization becomes less satisfying.[51] Russia is now experiencing the rapid urbanization which was seen in Britain a century ago, in Germany from the 1870's, and in the United States from the end of the nineteenth century, and this will affect its agriculture by increasing demand and making possible more specialization than before. Therefore the regional complexion must change. Modern pioneer efforts to make regional schemes in Britain began in the south-east, and showed a gratifying correlation between physical features and agriculture (see p. 64); but once London was reached workers like Unstead began to speak of it as itself a regional entity.[52] But though in an area like Greater London the existence of some 800 square miles of buildings, roads, parks, docks and the rest is as obviously a regional unit as a chalk downland, any geographical description of Greater London must include a consideration of its physical features, for these have influenced its growth from the very beginnings, affected its transport routes, given it rich potentialities as a port fully used from the nineteenth century, and provided sites for parks and suburbs such as heaths and commons surrounded by houses such as Hampstead, Wimbledon, Clapham, Blackheath and many more.

It is sometimes supposed that modern regional geography has little concern with physical features. Assuming that inevitably the work becomes more detailed with the passage of time, one can argue that the concern with physical features increases rather than diminishes. If, for example, a worker deals with a few square miles of the American cotton belt, he may find the key to varied cropping in the particular local drainage conditions; if a worker deals with some part of the European Mediterranean, he may find within a few square miles some fertile cropland, an area of horticulture, some vineyards on sunlit slopes, a pine forest on sandy soils, olive groves on hill-sides and a *maquis* on limestone or a heath with *ericas* and the like on exposed uplands. In the distribution of settlement, too, the availability of water becomes a matter of concern—most British students learn about springline villages at the foot of chalk downlands. In this connection, one may perhaps quote again G. G. Chisholm's statement that 'it is of the highest consequence to have a class of investigators whose constant and single aim is to see that the known causes that affect the value for man of place are never overlooked, and to be always searching for unknown causes that have the same effect'.[53]

The Problem of Regional Geography

Disappointment with the work of regional geographers has led many to wonder if the regional approach can ever be academically satisfying and to turn to specialization or some systematic branch of the subject such as geomorphology, climatology or economic geography. But to a large extent it was the appeal of regional geography that proved crucial in the modern growth of the subject: why then should there have been a change? There are many answers, but before embarking on some of them one would wish to say that systematic and regional geography need not be thought of as rivals but rather as complementary, for each can fructify the other. Of the older modern classics, none illustrates this better than Vidal de la Blache's *Human Geography*, in which the major principles are illustrated by local examples drawn from wide regional reading.[54] Modern economic geographies, and other works, show an equal debt to regional works as surely as some of these show a debt to applied economics. The mental process involved here is to establish some general principles or even— much more boldly—general laws based on local surveys. Any such principles must be constantly tested by more detailed survey: for example, it is broadly true that in the European Alps the limit of agricultural cropping and pasture runs higher on the south-facing slopes—the *adret*—than on those facing north—the *ubac*. Yet in a discriminating study of Alpine regional climatology, Alice Garnett has shown that the relationship between land use, slope and aspect is highly complex, for there are farmlands facing in every conceivable direction.[55] Further, some uncertainty exists on the actual incidence of plant growth in relation to intensity of light and heat.

Why then is so much regional geography disappointing? In the first place, much of it seems naïve—for example, Herbertson's natural regions, in effect climatic, no longer seem an adequate basis for further study though, like other world classifications, they were a great advance in their time. Their very success has been their undoing, for they have been adapted for classifications of increasing refinement showing, for example, the length of the growing season for various crops, the effect of weather and climate on people, the accumulated temperatures related to crop growth, the amount of water available for crops allowing for rainfall and evaporation. Equally many of the world or continental regional maps of physical features have proved to be merely a basis for more detailed work. But if one looks back a hundred years, the work of generalization has at least been a useful stage, though a warning[56] about its dangers appeared in America as long ago as 1857.

'Hypothetical geography has proceeded far enough in the United States. In no country has it been carried to such an extent, or been attended with more disastrous consequences. This pernicious system was commenced under the eminent auspices of Baron Humboldt who, from a few excursions into Mexico, attempted to figure the whole American continent . . . On the same kind of unsubstantial information maps of the whole continent have been produced and engraved in the highest style of art, and sent forth to receive the patronage of Congress and the applause of geographical societies at home and abroad, while the substantial contributors to accurate geography have seen their works pillaged and distorted, and themselves overlooked and forgotten.'

Secondly, many works on regional geography drag wearily through a sequence of apparently unrelated facts of physical features, climate, vegetation, agriculture, industries, population and the like, with little attention to the relationship between the physical environment and the inhabitants and—in some cases—they include the most involved digressions into such matters as the physical history of an area: indeed at one time some regional writers gave as their physical section a synoptic geological history. If, however, one deals with Denmark, an obvious starting-point is the variety of glacial drifts and other superficial deposits that give the country its varied but nowhere dramatic physical features, perhaps best explained in terms of the history of the retreat-stages of the Quaternary Ice Age; if one deals with the Netherlands, the reclamation of land from the sea gives an obvious key to much of the country's agriculture, and even to the form and modern growth of its towns. The trouble has perhaps been that many regional geographers have tried to include too much.

But there is a third, and more subtle difficulty. Perhaps due to the marked success of the *pays* treatment in the Paris basin, some geographers have written as if each area distinguished by their regional treatment has a quality of uniqueness, or at least an individuality, a personality all its own. It is of course true that there may be marked differences between the agriculture and standard of living between adjacent regions, but, as a recent trenchant critic has shown, there are other aspects of human personality expressed in art, drama, sport, religion, that may override differences in economic standards clearly related to the physical environment. Or are they clearly related? If the author may illustrate this point from his own experience, two environments he has studied, south-west Scotland and north-east Ireland, are physically closely comparable; but owing to differences in economic and social history in Scotland the farms are two to three times as large as those in Ireland and the Scottish rural population density only one half to one third that of Ireland.[57] It matters little if a geographer finds an explanation of something he studies in history—often historians find the geographical explanation to be crucial.

A recent tendency in regional writing is to take a theme and build the work around it. Preston James,[58] in his great pioneer work on Latin America, uses vegetation and major physical features with marked success to lead to a study of human occupation and population distribution. Work on land use in Britain has been based on the detailed mapping of the way each acre of the country is used, but it has led naturally to a consideration of physical features including climate and soils on the one hand, and to a study of the relation between town and country on the other. It may not claim to be a complete regional geography; but the very word 'complete' begs the question. And here lies an interesting opportunity, not perhaps adequately seized by geographers but at least appreciated by many modern planners, to find out everything possible of geographical, historical and economic interest about an area. Geographically this will involve not only a close study of its present landscape but an effort to reconstruct its past landscapes—an enterprise already followed by a number of historical geographers. Incidentally this is virtually what H. R. Mill suggested nearly sixty years ago (pp. 63–4).

Perhaps the most dangerous word in regional geography is 'natural'. It has so often been applied in a supposedly scientific sense to provide a world framework into which men and their activities must somehow be fitted. Even in such connections as the Mediterranean lands, it may mask by generalization endless local varieties of scene. But however one criticizes the underlying assumptions of regional geography, the fact remains that it has been a valuable, indeed an essential contribution to the understanding of the world. If one turns back to works of a hundred years ago, one finds lists of the political units with their towns and rivers, statements on the average elevation of a country with the proportion of its land between different levels, and such sayings as this, culled from a text of 1866: 'The varied character of the southwestern shores of Ireland deserves especial notice. Of its many inlets, the finest is Dingle Bay, which penetrates the land for upwards of thirty miles.' At least it is now possible for most countries of the world to find some literature published as regional geography which will give a synoptic view, even an analytical view, of the country's landscape. It may even be possible to find articles that for small areas give a closely-reasoned account of its physical environment and life, of how perhaps from one generation to another the limit of farming has been pushed farther up hill-sides, into common lands and heaths, or deeper into forests. Many would agree that there has been premature generalization, and the need may well be for more local study before new and more satisfying generalizations can be made. Equally one may hope that contributions will be made to regional geography by the systematic studies now increasingly practised, for example by authors who write on such aspects of economic geography as the agricultural life of small areas, or on detailed surveys of towns, or towns in relation to rural areas. As regional geography gave much to systematic studies in earlier years so now it is fair to expect some return. But the theme of these last few pages is quite simple—that much has been done by regional geographers, but only enough to make one wish for more.

CHAPTER SEVEN

ECONOMIC FACTORS IN GEOGRAPHY

'Commercial' and 'economic' geography; natural resources;
the use of resources; agricultural changes.

'Commercial' and 'Economic' Geography

PART of the modern interest in geography arose from its commercial applications, which have been clearly appreciated for the past century as transport has removed isolation from much of the world and made unknown lands spheres of influence and trade. Several of the world's geographical societies, including in Britain those of Scotland and Manchester, had commercial geography as a main source of attraction and some of the first university courses were also of a mainly commercial and economic character. The terms 'commercial' and 'economic' have acquired slightly different meanings:[1] in 1882, Götz suggested that 'economic' geography was to be more academic and commercial geography essentially practical. The tendency in university courses has been for 'commercial' courses to be of obviously clear use to the students drawn from schools of commerce, and for a deeper geographical framework to be laid in 'economic' courses, with more emphasis on such fundamentals as climate, physical features, the limits of particular crops, or even the historical effects of the modern opening-up of the world by transport. Much interesting work has been done on the past distribution of population and industry in Britain, and indeed elsewhere, and in Britain the need to conserve at least some of the old mills, water-wheels, factories, foundries, canals, railway stations, if not necessarily by preservation at least by photographic and other recording, has been recognized by the Council of British Archaeology,[2] whose work now includes industrial archaeology. The interest of folk-museums is partly economic, and such bodies as the National Trust of England and Wales have assumed responsibility for the care and preservations of various old mills and farm-houses.

Not inevitably, therefore, is economic geography directly

utilitarian in its approach to the world. Indeed Chisholm's great *Handbook of Commercial [sic] Geography* has many references to historical factors, and includes both a treatment of commodities and of countries: so inclusive is the author's work that virtually all the commodities on the market anywhere appear to be mentioned. But Chisholm was no determinist: indeed on the very first page[3] he notes that 'the great geographical fact on which commerce depends is that different parts of the world yield different products, or furnish the same products under unequally favourable conditions'. Commerce may increase the variety of commodities available at any place and also equalize, 'more or less, according to the facilities for transport, the advantages for obtaining any particular commodity in different places between which commerce is carried on'.[4] Transport is crucial, so crucial that one wonders why some geographers write of it as a factor only just discovered. Nowhere has the relevance of transport been more fully realized than in Russia, where the opening up of new agricultural lands and reserves of minerals has only been made possible by the provision of railways and, in some circumstances, waterways.

Russia's economic geography appears to be changing at a swift rate—even between 1939 and 1959 the population of the U.S.S.R. increased from 190·7 millions to 208·8 millions (1959 boundaries), in spite of enormous war losses. But a more significant aspect is the redistribution of this vast population, for in 1939 60·4 millions (32 per cent) were in towns, and in 1959 99·8 millions (48 per cent): meanwhile the rural population had declined by 21·2 millions. Inevitably, in order to understand so great a change, one must know something of the modern political organization of the U.S.S.R.: equally, one may add, study of the population distribution of China and Japan during the past hundred years is aided by some knowledge of the changing national policy of Japan and of both the Confucian ethos and the Communist changes in China. And it is also essential to inquire into the definition of a town in the U.S.S.R., as in any other country: as noted on pp. 119–21, administrative geography is often taken for granted. Even so, the modern rise of the towns in Russia is only an expression of the Industrial Revolution combined, here as elsewhere, with increased production from the land by a diminishing number of workers. At the 1851 Census of Britain it was regarded as of phenomenal importance that there were more people in the towns than in the country-side, 'a situation that had probably not existed before, in a great country, at any time in the world's history'.[5] Since then it has been seen in many countries, and will be seen in Russia almost at the time these words are printed.

Although by no means unique as an example of town growth, the Russian urbanization shows the immense enthusiasm with which it has been fostered. Frank Lorimer,[6] writing in 1945, said that the initial problems of the U.S.S.R. in the 1920's were three: first, an excessive dependence on agriculture at a low technical level; second, a retarded development of industry due partly to the lack of capital equipment and of skilled labour; and third, the inadequate economic integration of different regions in a country largely underpopulated, with high transport costs. The Soviet change has apparently stimulated economic geographers into following directly practical lines of work: V. C. Finch,[7] writing in 1944, noted that Russian 'economic geographers are no longer interested in the bare statistical inquiries of pre-revolutionary economic geography, but are specializing in intensive studies of the present regional distribution of types of economic activity and in devising principles and practical programmes for a co-ordinated development of these resources in terms of regional integration'. In short, a vast territory is now experiencing an intensive industrial and agricultural revolution, in which geographers have found practical scope for their research, publication and mapping, some of them in planning offices. The war period from 1941–5 not only brought immense devastation with the opportunity of later reconstruction, but also strengthened a tendency already seen—movement eastwards into the Urals, Siberia and Transcaucasia, of population, industry and mechanized agriculture.

Basic to all study of economic geography is the distribution of population which is continually changing. If for simplicity such study is regarded as of man at work, the concern is with the sower sowing his seed, the factory hand turning the lathe, the scholar reading in his study, the clerk filling in his forms, the actor living his lines. Along such a line of argument, economic geography is closely linked to the social aspects of the subject, which are discussed in Chapter Eight: equally, as shown on pp. 127–37 consideration of men at work, through an area's distinctive production, has led some authors, mainly but not exclusively Americans, to use economic distributions as a basis for regionalization. And the essential challenge to the student of economic geography lies in the constancy of change, which involves the necessity of keeping some historical perspective. In the present world, the fear of over-population nags at many minds as it did a century or more ago when there were only about half the present number of inhabitants on the globe. Although, in a deplorable modern catchword we are assured of certain classes of the community that 'they've never had it so good', the fact remains that a substantial proportion of the world's population, variously estimated but rarely at less than half, are underfed. Nor is that the only social problem. Not only are large sections of the world's population living in permanent and grinding poverty, but many millions are living in housing conditions prejudicial to health.

Economic geography at the present time needs much study of contemporary changes as well as an historical approach. The current changes in much of the world are so swift that it is hard to keep pace with them: perhaps for this reason one geographer commented, 'Economic geography is always out of date.' But it must be significant that in Britain and several other countries of Europe agricultural production increases annually though the workers on the land diminish, and it is equally significant that in the United States the losses of farm-land to non-agricultural use has been estimated to be as much as a million acres a year; while in Russia, according to reports, many times this amount of land is being added to the farm-lands annually.[8] Possibly these trends show that the three countries just mentioned are at different stages of economic development, but for the moment it is enough to say that their use of land is far from uniform. Indeed it appears to be governed, partially at least, by national economic and social policies; and the trend of the past thirty years at least has diverged from the fundamental premises of Chisholm's day, that each part of the world should ideally produce the crops to which it is most suited. National policy may dictate such enterprises as the extension of wheat and other cereal crop-growing far into the *taïga* (coniferous forest) belt or on to the steppes of Kazakhstan where the farms have to contend with early and late frosts, variable rainfall, and soils of moderate quality with a limited capacity for chemical fertilization,[9] to say nothing of such difficulties as transport. As the town population of Russia increases, so does its need for home-grown grain, and the former contribution of the Ukraine to Europe's wheat supply is no longer made. Many such examples could be given, especially as the governments of various countries in varying degrees influence or even control their imports and exports: to an increasing extent states plan their economies by fostering particular types of industry and by subsidies, propaganda and other means, they control, or at least influence, agricultural land use. The word 'use' also involves 'misuse'. The somewhat outworn controversy between the possibilists and the determinists should not obscure the fact that large areas of land have been ruined, in some places permanently, by efforts to grow crops or pasture stock in areas unsuited to such activity.

In recent years, economic geography has been re-vivified by a certain amount of polemic and argument and at times hindered by efforts at pontification. On the international scale, the issue is argued between Communism and other forms of political organiza-tion as ways of arranging the agricultural and industrial life of nations: dictatorships are not infrequently armed with consider-able economic powers, varyingly used. But in the democracies the main argument rests on the extent to which the planning of pro-duction or even of the use of the land should be controlled at all. There are, for example, very marked cleavages of opinion in Britain on the need for government influence in the location of industry or the allocation of agricultural land for town growth. Equally, there are many views on the wisdom of using land of particular types for sheep-farming or forestry and at times passions run high on such subjects. But the advantage of such work as the Land Utilization Survey of the 1930's is that it provides a clear factual basis for arguments that are partially economic, but only partially as they involve such questions as social welfare and the preservation of amenity.

In 1937, Sir Josiah C. (Lord) Stamp (1880–1941) gave a careful survey of the relation of economic geography to general economic theory.[10] He apparently agreed with Chisholm that there was 'a tendency towards the equalization of economic development throughout the world in capital, in population density and in skill'. He used geographical examples to illustrate economic explanations of classes of facts, known as simple static or inductive static. In the first, or simple static class, one fact or set of facts is explained by another—for example, Antwerp as the nearest great port to the main manufacturing areas of Germany, has grown with the in-dustrial growth of Germany. If numerous similar facts emerge, the class becomes the inductive static, permitting a measure of generalization: study of Belgium, Holland and England, for example, reveals that all have a diversified industry partly, indeed largely, dependent on imported raw materials, with an agricultural life marked by high productivity and a strong scientific tradition. To a great extent this view is limited in time: change is considered more in a further class of facts, called by Stamp simple direct dynamic, which involves two sets of changing factors, such as the westward movement of slaughtering and meat-packing stations with the advance in the frontier of ranching in, for example, the United States or the Argentine. Similar instances can be given for other areas, such as the development of new processing factories for soya beans—said to be an industrial raw material of a thousand uses—as Manchuria has been steadily settled by Chinese farmers, penetrated by railways and given factories with capital brought in by the Japanese and Russians as well as the Chinese. Not all the examples given here are those of the original Stamp article, which will repay careful study: as shown on pp. 156–64 there are endless complexities in the location of industry.

In Lord Stamp's view economic theory is not based on static conditions and 'only as geography registers change over time can it be of full advantage to economic theory'.[11] One may justly ask whether any area of the world retains for long the same combination of advantages for the production of agricultural or industrial goods. The cotton area based on Manchester as its main market has lost so heavily in its export trade since the early 1920's that the industry now has hardly one third of the workers it once employed, and widespread unemployment has been prevented only by the growth of other industries and by outward migration. In Cheshire, the dairying area still remains comparable in extent to what it was before 1939, but with the difference that the milk is now sold, on most farms, to the Milk Marketing Board or to co-operatives and commercial firms: cheese-making as a farm craft has almost— happily not quite—ceased.[12] Indeed within the same county there are many fascinating examples of industrial change, such as the replacement of the salt industry at Northwich by chemicals. And Crewe developed beside its six-point railway star, made there— it appears—because land could not be acquired around Nantwich, the historic road centre four miles to the south. In this case the availability of land near a railway crossing place provided circumstances favourable for workshops which were seized at the right time, as in Swindon.

Some authors have expressed themselves forcibly on the advantages to mankind that may be expected to follow from commerce: of these one may quote Chisholm,[13] writing in 1923, of the 'goal of commerce (as) . . . that stage in . . . evolution . . . when the inhabitants of the earth will be able to enjoy the greatest possible variety of commodities, supplied at the least cost and with the greatest attainable stability of prices'. This depended, he thought, on three things: first, the completion of all the main lines required in the network of communications; second, the education of all the peoples of the earth, 'intellectually and morally, as nearly as possible to the same levels'; and third, the discovery of new sources of power. The time would come when 'coal, ores and other irreplaceable materials are economically exhausted and we are thrown back . . . on direct sun-heat'. This last point reads rather quaintly, though attention has been drawn at times in recent years to the rapid rate at which raw materials have been used in the twentieth century; and had Chisholm been writing forty years later he would undoubtedly have spoken of atomic power. To him, the world disparities in standard of living appeared pernicious: for example, low wages in the East made for unfair competition in overseas markets and kept the home producers in poverty. He used Lord Beveridge's definition of optimum density of population[14] as that 'which bring the largest return per head of the population'. Food, shelter and fuel were basic requirements, and the aim of commerce was to make the surplus as large as possible—even in Great Britain large sections of the population had too little milk. Chisholm thought that in Japan, China and India large numbers should be removed from the land, and he finally works round to the view that the 'world saturation' of population might bring 'more wars, more famine, more disease'. Like other writers, he was clearly conscious of the law of diminishing returns in agriculture: in an earlier article, he had asked how far the yield per acre could be increased and had noted[15] that 'it is one thing to increase the production of wheat from nineteen to twenty-eight bushels per acre and quite another . . . from twenty-eight to thirty-seven bushels'.

This preliminary discussion has perhaps indicated some of the problems with which economic geography deals: from a somewhat grim survey of products, it has become a subject that holds at least one key to the study of population problems of wide social significance. Nevertheless it is fundamentally concerned with the resources of the world, industrial and agricultural, and now, as in the first heroic days of the early geographical societies, the aim is to find what resources exist. But that is not all, for the study leads on to the distribution of industry and agriculture and so to that of population. It reveals a changing world, indeed a world changing so fast that the 'current' 'up-to-date' economic geography of today may be the historical economic text of tomorrow. In the remainder of this chapter, attention is given to four main themes: natural resources, their present development, the location of industry, and agriculture.

Natural Resources

An American geographer, H. H. McCarty, has said that economic geography is becoming the branch of human knowledge whose function is to account for the location of economic activities on the various portions of the earth's surface.[16] An obvious line of first inquiry is the intrinsic natural resources, but these do not of necessity give any clue to the economic activities at any place. Great Britain, for example, imports much of its raw materials for its manufactures and a substantial part of the food to feed its population. Japan, during the past century, has built up an economy based on world trading after living for many centuries in self-sufficiency. Such developments rest on the application of modern technology to industry, the skilful management of trading, the efficient use of power and the provision of modern transport. Of these only the power may be locally available at the outset, for all the others may be brought to an area from outside sources: in Britain and Japan a clear national need is the maintenance of a large overseas market in which to buy and sell, as in neither case could the economic activities be explained solely on a basis of natural resources.

Reconnaissance geography has a romantic and adventurous sound, but it is only a first stage in a regional survey that may reveal new mineral resources, agricultural potentialities or exploitable forests. Geological prospecting may be a social service. As recently as 1849 the Cleveland iron ore field of Yorkshire was discovered—or probably rediscovered, as it was apparently known to the Romans. Every discovery by geological prospecting of new mineral resources ('Boys, it's oil') has its dramatic appeal, and modern Russian works on geography have accounts of new sources of oil, coal and other minerals, some of them in places remote from modern transport. So far there can be no final assessment of the resources of the world in minerals, for some are little explored, some perhaps undiscovered, and others too remote at present to be commercially successful. But mineral resources differ from those of forest and field in their liability to exhaustion, and some writers are already showing concern at the present world rate of use: indeed, some estimates are that during the first half of the twentieth century more was removed from the earth than during all previous time.

Forests have been removed from the land ever since prehistoric times for a variety of reasons. The most primitive forms of agriculture, still practised in parts of the world, consist of burning a patch of forest and sowing seeds on the naturally-enriched land left behind, cultivating this for a few years, then moving on elsewhere. In China forests were burned to remove wild animals, and under conditions of heavy rainfall soil erosion followed so that in some places the hill-side became useless for any economic purpose. But under modern conditions a forest may be a valuable and scientifically-managed source of constructional timber, newsprint, and an increasing range of manufactured commodities. The main timber reserve of the world lies in the *taïga*, or northern coniferous forest belt, in Scandinavia, the U.S.S.R. and Canada: there are still vast areas available and the normal practice is to clear trees by thinning and to allow for natural regeneration. Indeed, as shown on p. 157, there are areas within the *taïga* that are unlikely to be used for some time, if at all, owing to the prohibitive cost of transport.

Agricultural resources in any area require delicate assessment: on the one hand, many parts of the world have produced far more than was ever thought possible, and on the other many areas of pioneer settlement have been abandoned. At present one can only watch with interest the claims of the Russians that vast areas, some of them equal to Great Britain in size, are reclaimed for farming in a few years and that further great expansions may be expected. But the experience of the New World may give a warning: the Great Plains of North America were almost entirely neglected by the Indians, yet part of it became a world granary, except where farming was pushed beyond its safe limits and the fertile soil was removed by erosion. Some such areas could have been used as grasslands, but overgrazing was harmful as in times of drought animals pulled up and devoured the stems and roots of the grasses, leaving the soil open to erosion in dry periods. In many parts of the world there are indications of withdrawal from marginal areas though it is equally certain that there are other areas that could be used for agriculture, especially if irrigation is possible. In the world quest for food, such countries as Brazil come to mind but, as Preston James[17] has shown, the expansion of the area used for agriculture involves three things—the attraction of suitable migrants, land suited to the kind of agriculture they can practise, and the provision of transport and other services.

Nagging at the minds of thoughtful people everywhere is the fear that the population of the world may increase beyond the earth's capacity to support it, either in food or in the raw materials of industry. The warnings of Malthus (1766–1834) in 1798 were given when the population of the world was less than half, and probably only one-third, what it is now; but at all times the crux of the problem has not been the relation between population and total natural resources, but rather the existing development of natural resources. To most people it is a source of shame that a substantial proportion of humanity should be permanently under-nourished, but this is certainly no new thing; rather has the dis-closure of a world problem been made possible by modern methods of dietetic analysis, and their publication by such bodies as the Food and Agriculture Organization of the United Nations. [18] And periodic disasters such as the Bengal famine of 1943 due to the poor local harvest of rice and the impossibility of importing supplies from Burma and Siam in wartime, show how precarious the hold of large populations may be on traditional food supplies.[19] On the other hand, considerable areas of the United States are at present reverting to rough pasture or subspontaneous woodland as pleasure grounds for city workers who buy up old farm-houses with land as country homes.[20] The British people are deeply conscious of the need to grow as much food as possible within their homeland, and the extension of towns at the expense of rural land invariably raises controversy, in many cases fought out before legal tribunals.[21] But the American view of land is different: it is an expendable resource, existing in vast quantity and available for industrial and residential growth; no doubt there will, in time, be a movement to conserve land for agricultural use.

Put simply, the issue is this. There is no clear estimate of the total food-producing capacity of the world at present. As Sir John Russell[22] has said in his book, *World Population and World Food Supplies*, 'less than 10 per cent of the world's land surface is cultivated; ways can still be found of expanding into the 90 per cent at present untilled'. But those who read his book will discover that agricultural expansion is not easy in many countries, for political and economic reasons as well as the inherent difficulty of using various soils: the Dust Bowl tragedy of the United States in the 1930's or the British-sponsored ground nuts scheme in East Africa act as cautionary tales for those who go forth inadequately armed with a basic knowledge of soils, vegetation and rainfall—and especially rainfall variability. Nevertheless, Sir John Russell's main point stands, not least because there is the hope of a wide-spread dissemination of scientific knowledge in farming, combined with more efficient plant breeding. In every country the standard of farming differs widely: areas apparently homogeneous in soil and climate may have a wide variation in efficiency from one holding to the next, as the writer found when, as a student, he first took out a map to study land use. But this very disparity gives ground for hope. Sir John Russell[23] believes that 'the picture that finally emerges is one of tempered optimism. In all the countries examined, there is a considerable gap between the best and the average food producer which can certainly be narrowed, thereby increasing the output of food'. Equally the possible effects of scientific knowledge must be considerable, even though they percolate through farming communities only slowly.

For the non-agricultural resources, it is equally difficult to forecast future needs. In Britain and Ireland there are many who wish to extend the area forested, especially in the mountain ranges, not only to reduce imports but also to give employment and to ensure supplies of an essential raw material, especially in times of war. But there is a conflict of views, as others wish to preserve to the maximum possible extent the use of mountains for pastoral farming, having an independent agricultural community rather than a group of forestry workers living in a neat village near a sawmill. Through the *taïga* belt of northern Europe there are still possibilities of increased agricultural settlement, located in some places on alluvial river plains, drained marshes or peat bogs rather than forest clearings. Not unusually, men settled on the land work part-time in lumbering or in sawmills. For much of human history wood has been used both as fuel and for constructional purposes: more and more, though retaining its use for building, furniture and allied uses, it has become a raw material for many derived industries.

The Use of Resources

Mineral exploitation, said Brunhes,[24] 'fixes man's labour, suddenly and for the time being only, at one particular spot on the earth.' This problem is seen in its most extreme form in the mining settlements in deserts or similarly inhospitable regions, but exists everywhere where extractive industry is carried on. In Britain the coalfields were magnets for immigrants during the nineteenth and early twentieth centuries, but many of the mines are now exhausted and some coalfields have seen decreases of population as dramatic as the increases of former years. Theoretically, it might be reasonable to regard any mining area as transitory and to provide houses, shops, schools and other social services only for the estimated life of the mine. But in Britain, though miners have the tradition of moving to other areas in search of work, either daily or as permanent residents, opinion generally favours the special provision of factory employment in mining areas such as South Wales, Durham and Northumberland, or West Cumberland, all of which are development areas (p. 160). True, there are voices, including those of some distinguished economists, which are raised against this practice; but the British are reluctant to see such areas abandoned to continuing economic decay.

Other factors than mere exhaustion enter into the precarious history of mining. In Britain there has been a reduction in the sale of coal recently, due in part to the increasing use of oil for heating; the earlier reduction in the demand for coal, during the 1920's, was caused by the loss of export markets. But the rapid world consumption of oil raises questions: it may be, however, that within the next few decades much of the world's energy will be acquired by atomic power. There is no certainty in the future of any mineral working: many landscapes are strewn with the relics of past mining activity, abandoned not because of exhaustion but because other sources became cheaper—in Cornwall, for example, copper mining has ceased, tin mining is practically extinct, but kaolin (china clay) extraction is prosperous, both for the home and overseas market. But of all stories of mineral ores, the classic is the use of the *minette* or second-grade ores of Lorraine which from 1879 became valuable, as by the Gilchrist-Thomas process the phosphoric content could be extracted to provide valuable agricultural fertilizers.[25] The use of mineral resources therefore has shown, and may well show in the future, quite unpredictable trends, not least through the development of synthetic products in the chemical industries.

The possession of natural resources is no guarantee of their use. This point has been frequently made, and one example can be given: in 1951, W. J. Eiteman and Alice B. Smuts wrote an article 'Alaska, Land of Opportunity, Limited', with the object of removing a misapprehension that millions could find a home there.[26] True, those who went would not have to contend with frontiersman hardships, as they could arrive by streamlined planes or luxurious steamships, stay in good hotels, find buses, taxis, supermarkets, beauty parlours and cinemas. But the economic limitations are such that the authors suggest that even a population of 'two or three hundred thousand people should be classified as extremely optimistic'. The Russians owned Alaska from 1741 to 1867 and made great profits from the furs of the sea-otter: they never knew of the gold and copper resources, found in 1896 and 1898. Gold mining and fur trapping still survive, but the main industry of the present day is the canning of salmon, of which Alaska provides seven-eighths of the world supply. But these industries show little prospect of immediate expansion, and hopes of a development of iron ore mining and newsprint supply are hampered by distance from consumers: even if the Alaskan forests could supply one-third of the U.S. needs, they are unlikely to do so as long as nearer forests, including those of Canada, are available. Comparisons with Sweden have been made, yet within 700 miles of Stockholm there are 100,000,000 people, within 700 miles of Ketchikan there are less than a million people. There is the permanent problem of distance from markets, as air transport is expensive and road hauls long: the natural resources are not at present economic resources.

It is quite untrue to say that distance has been annihilated and that, given the natural resources, there is nothing to prevent the development of appropriate industries. Even so, many resources are developed in extremely remote places, of which an example is the iron ore mines of northern Sweden, at Gellivara and Kiruna, where several thousand people live under a *tundra* climatic régime. But here the Luleå-Narvik railway provides an outlet, and the ores are shipped from Luleå in summer and from Narvik in winter. Many examples could be drawn from Scandinavia of industries that owe their position to some local advantage of site, in some cases power supply. At Höyanger, on an arm of the Sogne fiord, an aluminium plant was erected from 1916 and a town built for 3,500 people: the bauxite comes from Greece, but originally from Provence, cryolite from Greenland, petroleum coke from the U.S.A. and coal from several European countries.[27] There are many other examples of industries on Norwegian coasts, using water power—the only local resource, drawing both their raw materials from afar and sending out their products by sea. Although there is in Scandinavia, especially in Sweden, a long tradition of industry, its modern development rests largely on water power and dates mainly from the 1880's.

Local resources favour the growth of numerous industries: in Britain, for example, iron working developed where ores and suitable coals were close together, for example at Merthyr Tydfil and other places in the South Wales coalfield. Many works associated with local natural resources failed to survive their exhaustion, but not invariably, as the continued success of the Consett works in Co. Durham shows; even at Stoke-on-Trent, one ironworks still remains. Two modern steelworks on the Jurassic ore belt of England, at Corby and Scunthorpe, depend on coal transported from other fields; at Sheffield and Rotherham, with all the local ore long since exhausted and much of the coal brought from a distance, the steel industry is apparently impregnable. Initially associated with some natural resource, an industry may survive through its efficiency and organization. The Pennines and Rossendales became the greatest cotton-producing district of the world during the nineteenth century, having plentiful supplies of soft water for use in the washing, bleaching and dyeing processes, and streams which could be harnessed to drive water-wheels. Later, with steam power, there was coal available locally. During the latter part of the nineteenth century, the fear was constantly expressed that raw materials from America, Egypt and India might be inadequate, but few saw that the real threat was the growth of cotton factories competitive with those of Lancashire.[28] When chemical means of softening water were discovered, the initial locational advantages of the cotton industry had gone, but it survived in the district, partly because it had acquired a strong commercial organization centred in Manchester.

As a broad generalization, one could say that many industries owe their location to some natural resource, which may be mineral deposits, power or water. The search for industries tied to raw materials is not altogether easy, but some manufactures based on agricultural raw materials show a certain degree of correlation.

Co-operative creameries for butter-making appeared in Denmark during the 1870's, and shortly afterwards in the other Scandinavian countries, in Holland and Ireland: they were located at intervals of a few miles as the farmers carried their milk to them daily, with horse transport, and took home the whey for pig-feeding. Industry was effectively carried to the countryside, but the bacon factories were less numerous as the sales of pigs were infrequent. Equally some modern canning factories are located in areas where good supplies may be expected, and not unusually the existence of a canning factory provides a stimulus to vegetable growing in the vicinity. But the direct correlation of an industry with its raw material is precarious. Irish linen manufacture began as a domestic trade with the flax grown locally, prepared in a local scutching mill and finally spun in the farm-house for sale in the Cloth Halls: from c. 1825 to 1865, the factories gradually acquired the trade and as early as the 1840's there were complaints that the local supplies of flax were inadequate, so that in some years as much as half the raw materials had to be imported, largely from the Baltic provinces of Russia. That flax could be grown successfully was abundantly demonstrated during the world wars of the twentieth century, but the acreage declined from c. 120,000 in the 1939–45 war to only forty, the lowest ever known, by 1959. Of the agricultural industries named, there must be abundant local supplies of milk for butter, pigs for bacon, and vegetables for canning and deep-frozen foods; but linen manufacture resembles cotton in its survival on imported raw materials, largely because the raw material is only a small part of the final cost of the article. And the spinning and weaving mills of Northern Ireland have successfully used synthetic fibres, such as nylon, rayon and terylene, all of which are now manufactured locally with imported raw materials.

One could multiply examples for ever; and any generalizations on the geographical influences on industrial location must depend on the study of particular cases. The success or failure of an industry does not and cannot depend on geographical factors only, for there are numerous economic and historical aspects of industrial location; allowance has also to be made for the deliberate planning policies of governments for social and political purposes. For reasons of strategy or even prestige, a country may wish to have its heavy industries, its factories producing essential foodstuffs, its firms producing consumer goods or even luxury articles that will sell in export markets or strengthen the tourist appeal. Similarly, a country may favour one type of industry against another: of this the most obvious example is the Russian concentration on heavy industries rather than on consumer goods during successive five-year plans. And social policy of a very different kind is seen in efforts to carry industries to workers, as for example in the development areas of Britain. Can one recognize in this a feeling of responsibility for past services? The idea in some minds is that the people of the development areas, which include several of the historic coal and iron areas of Britain, did much in the past for the nation by their labour, and should therefore be helped now. On the other hand, some think that a more logical approach is to accept the view that under modern conditions industry is attracted to the Greater London area and the West Midlands and to avoid the artificial stimulation of areas that may have served their day.

No simple correlation between industry and raw materials exists, yet certain conclusions are possible. Firstly, an agricultural industry, such as a butter creamery, a cannery, a bacon factory, a brewery or distillery may encourage local farmers to provide some profitable raw material: milk for the creameries, vegetables for the cannery, pigs for the bacon factory, barley for the maltings or direct supply to breweries and distilleries, are obvious examples. Yet in some cases factories of these types find the supply of raw materials difficult; at no time, for example, were creameries successful in those parts of Ireland where cattle-raising rather than dairy-farming was traditional.

Secondly, an industry owing its initial location to local raw materials or power may survive when these natural advantages no longer exist. The linen trade of Northern Ireland now lives on imported flax; the steel industry of Sheffield on ore from other parts of Britain and foreign suppliers—even its knife-makers no longer use the Pennine streams but electricity from the national grid; the cotton industry of Lancashire and north-east Cheshire gathers its supplies from the world and has contracted through foreign competition and not—as some nineteenth-century writers feared it might—through any failure to acquire raw cotton. But it still survives in its original area, even though it no longer needs the soft water and power of Pennine streams now that water can be chemically softened and power made available anywhere. Thirdly, an industry may survive, change in character, and acquire new raw materials which may be processed locally. The linen trade of Northern Ireland has greatly changed in character during the past thirty years. Always having, in damask weaving, a tradition of weaving silk with linen fibres, it now uses a vast amount of the newer synthetic products, manufactured to an increasing extent in Northern Ireland which, like the British development areas, ardently desires to attract new industries as students of its propaganda will realize. The metamorphosis which has transformed some motor factories into aircraft producers or, in recent economic history, cycle manufacturing into motor-car works, is another clear example.

An eighteenth-century historian, William Hutton,[29] said that in the Black Country some industries 'spring up with the expedition of a blade of grass and, like that, wither in the summer'. But the Black Country, with Birmingham, has always had new industries based largely on iron and steel, even though as early as the eighteenth century the whole area did not support itself in pig iron. Why? Explanations based solely on natural resources, transport, power and other factors favourable to industry, may still be unconvincing. Even less convincing perhaps as compelling factors are arguments such as climate (so frequently adduced), the social or religious outlook of the people, the local banking system, the availability of labour, a tradition of industrial peace, a large extent of flat land close to rivers, canals, railways or roads. Yet any one of these, or a combination of them, may be helpful. Nor are these all the aspects concerned—in some cases, the propinquity of firms able to supply component parts, the existence of good marketing through exchanges and mercantile firms, the ease of acquiring imported raw materials, may be influential. Above all, under modern conditions, the demands of vast populations grouped in large towns and conurbations affect industrial location: it is, for example, no accident that many of the capital cities of the world, even including Moscow, have a number of consumer goods industries, many of which depend not only on local sales but also on the great distributing wholesalers and on the raw material suppliers prominent in great cities.

In any attempt to look backwards, one is impressed by the extent of change in the industries of particular towns or areas. In this there may appear to be an element of chance, which for example led to the outskirts of Oldham, Lancashire, a manufacturer of electrical goods looking for cheap premises: from this sprang a great concern with several thousand workers. Equally the vast expansion of the motor-car industry at Oxford has its roots in a small bicycle shop that existed some fifty years ago. Almost any town history will show that places have run through a gamut of industries, losing some and perhaps gaining others for a variety of reasons. But of all the influences on industrial location, none is more relevant than transport. And this was not provided on a basis of altruism: when the railways of Britain were built, the first aim was to connect London with Birmingham, Manchester and Liverpool, the major towns of the time. And when the network of lines spread across the entire country, chiefly during the 1840's, town populations rapidly grew through the widespread industrialization.

The German writer, A. Lösch (1906–45) has drawn attention to the widespread spacing of industry in the United States, and shows that 60 per cent of all production was consumed in the state where the manufacture was carried on: if certain categories were excluded, such as shoes, women's clothing, furs and jewellery, car parts, tobacco, wood products and a few others, 70 per cent was sold 'locally'—that is within the same state. Six commodities had over four-fifths of their produce consumed locally—artificial stone, ice-cream, wood from sawmills, bread, beverages and manufactured ice. Lösch[30] notes that there is an abundant literature on industries concentrated in particular areas, such as iron, steel, chemicals, machinery, glass, clocks, mining or shipbuilding, and comparatively little on those dispersed more widely through a whole country, such as those named above, with others such as brickmaking. Much of his data, however, refers to the year 1929, since when new techniques have developed. In Britain the past thirty years have seen many changes—for example, in some areas bread circulates over a wide area through modern methods of transport, and ice-cream manufacture has been concentrated through deep-freeze techniques. On the other hand, some industries—such as brewing, bottling from casks, and mineral water production—are widespread, as bottle-breakage costs figure largely in the cost. Lösch also discusses the changing location of industry, such as the shift of the American cotton-spinning to the south, the movement of iron and steel production towards ore deposits, both in U.S.A. and in Germany. Transport costs enter deeply into the cost structure but not necessarily in direct relation with distance— for example in Germany coal freight prices were subsidized for long distances, partly to avoid the excessive concentration of industry on or near coalfields.[31] And in Canada, the transport of some goods from Montreal to Calgary (2,240 miles) costs one-third more than to Vancouver (2,880 miles) because the latter rate had to be competitive with water transport.

One fascinating feature of Lösch's work is the study of prices and costs which differ considerably from one area to another. It appears that Lösch had in preparation a book on geography and prices,[32] which never appeared, though some of its conclusions were used in *The Economics of Location*. The Canadian example given above shows the relative cheapness of water transport. Some industries of the Atlantic coasts with rivals in the Mississippi area could not compete in the Middle West because of high railway costs, yet they were competitive on the Pacific coasts as goods were sent by sea.[33] New York pianos had a market area which extended for 500 miles from the Atlantic coast and 1,000 miles from the Pacific coast; of course, nothing will prevent the really selective purchaser with money from buying what he desires regardless of cost. Much can be done in economic geography by inquiry into the market areas of particular commodities, which are constantly subject to change: it is clear that Lösch, who died in 1945, had no intention of trying to set up a series of general laws and the author of the memorial preface says that 'the highest praise of his book will be if in the future it can be said that the work it has stimulated has made it obsolete'.[34] When that happens, the work of such authors as Lösch becomes of historical significance: all work on economic geography will gain, not lose, by precise dating, especially when statistical material is used.

In a country like Great Britain, maritime trade has long been prominent, and therefore the rise of port industries during the past century or more has been considerable. It is not hard to find examples: one of the most interesting is Cardiff, where imported iron ore and scrap meets coal at the dockside, or more recently Margam, near Port Talbot, with a similar combination of raw materials. Even so, social pressure in the 1930's was partly responsible for the erection of a blast furnace at Ebbw Vale in a mining valley of Monmouthshire. But the general trend, once the ore was exhausted in the South Wales coalfield, was to bring the coal to the port to meet the ore. At Middlesbrough, Yorkshire, coal came from Durham to the north and ore from the Cleveland field to the south (now a tiny fraction of the whole supply), from the Midlands ore field and from overseas; but in an earlier phase, iron smelting was widespread in the Durham coalfield. Much depends on the facilities of a port: the Manchester Ship Canal, built by 1894, has become what some of its advocates hope, a magnet for industry, which includes a trading estate (1896) at Trafford Park with over 50,000 workers, a steel plant at Irlam drawing its ore from ships coming to its own wharves, and oil industries near Manchester and at Stanlow near Ellesmere Port. But the canal has never been what the Liverpool cotton merchants feared it might be—the main route for cotton imports to Manchester. One could multiply examples, but mention must be made of London, recently treated in an admirable short book by J. Bird,[35] who shows that at present the effective economic limit, for all industries except petroleum refinery, is at Tilbury and Gravesend, twenty-six miles downstream from the city centre; he adds, however, that developments are likely east of this point.

An endless variety of special problems arise in economic geography, and the general trend of inquiry has been towards the local study, or at least the local example. It would be, indeed it has been, extremely dangerous to try to establish general principles, still more general laws, before more detailed local studies had been made. On pp. 91–2 it was noted that Chisholm was deeply aware of the unanswered, and in many cases still unanswerable, questions that arise in economic geography. Not infrequently, valuable material has been brought into economic geography by regional study and it is there that many of the most effective contributions will be made; the student of economic geography is confronted with constantly changing circumstances, with trends of industrial change that may involve widespread movements of population. In England at least, some modern problems of industrial location have become matters of social controversy, resulting in planning legislation: some of these are discussed on pp. 158, 160.

Agricultural Changes

Over a century ago, Sir Robert Kane wrote a book on the industrial resources of Ireland.[36] Himself a chemist, he had great hopes of raising the fertility of the land by artificial manures as well as by the known and tried methods of adding lime, marl or sand; he even gives analyses of soils. As Europe's industries advanced, so too did its agriculture; and 'improvement' was the catchword of the day. Nevertheless from the fall of the Roman Empire to the eighteenth century there had been little advance in European agriculture, though during the second quarter of the eighteenth century there were considerable improvements in England which included the development of new four-course (Norfolk) rotations, the keeping of stock for manure and the doubling of the wheat yield from ten to twenty bushels per acre. But this was almost unique, for in much of continental Europe the agricultural revolution dates only from the nineteenth century, partly through the use of phosphates, basic slag and Chilean nitrates from the middle of the century. At the beginning of the century, Germany's yield of wheat per acre was ten bushels, but it was thirty bushels by 1906 (and even more in Britain). Meanwhile, the settlement of new lands, not only in the United States and Canada, but also in Russia, brought new supplies of wheat to the world market; and even within Europe the agricultural area was increased by at least one-fifth, and probably considerably more, during the nineteenth century.[37] In other words, the world industrial revolution was accompanied by an agricultural revolution.

Whether at any time in history the world's population has been adequately fed, no one can say. And it is not original to comment that a better distribution of food is needed, both to enable food once grown to find a market and to distribute it to places where it may be sold: for example, until better transport is provided in countries like China, periods of crop failure may render millions of people beyond the aid of organized relief, and still less able to find employment which will enable them to buy food. This is no new problem, but current anxieties centre around the known world population increase of at least 1 per cent per year. As the American geographer, O. E. Baker, said in 1921,[38] 'the waves of population are beating against the barriers of adverse physical conditions all along the shoreline of settlement . . . As the pressure of population upon the agricultural resources of the nation rises, as land becomes scarcer and more valuable, greater and greater care must be exercised in using each kind of land for the purpose most favoured by the physical conditions . . . the growing of wood will enter into competition with other crops, especially for the use of the poorer lands.' And this is still true. As Baker pointed out in the same paper, some areas of poor soil in New Hampshire had been cultivated but later used for forests, which released farmers from a desperate struggle with poor soils, and gave to forestry workers a better financial return than that gained by the farmers they replaced.

To think of land reclamation as the inevitable solution of world food needs is dangerous. Baker showed that in 1920, Michigan had ten million acres, and Wisconsin and Minnesota another five million acres of land which could not be profitably used for crops but should be forested. Some of this land had been cleared, perhaps at heavy expense, farmed for a few years and then abandoned.[39] And in California, from 1889 to 1909, the acreage of wheat decreased from 2,683,000 acres to 478,000 acres; according to local reports, the crop 'wore out the soil', and one farmer said 'it was wheat, then barley, then rye, then good-bye'. Two-thirds of the former wheat land reverted to pasture or remained unused. By 1920, there were signs of increasing local specialization in American agriculture; in the semi-arid regions, the moister land was more intensively used, in some cases for fruit with irrigation, and the poorer tracts were kept in pasture or other light uses; two-fifths of the apples came from irrigated valleys of the western states and most of the rest from commercial orchards in small areas of Missouri, western Michigan, western New York, Virginia and adjacent states. 'It is not inconceivable,' Baker[40] comments, 'that this development in the United States may result eventually in that geographical differentiation of agricultural production so characteristic of portions of western Europe.' And in some cases the area of production has contracted rather than expanded: at one time cotton was grown as far north as Washington, D.C. and St Louis, Missouri, and in 1920 it was still grown in a few isolated valleys of the Kentucky mountains to mix with wool in homespuns; but once the crop became commercialized, it was restricted to areas where the mean summer temperature was 77° F. with at least 200 frost-free days. More recently, using modern climatic data, L. Curry [41] shows that within the 200-days limit the cotton-growing areas must have an annual average potential evapotranspiration* of thirty-three inches or more.

In short, crops are increasingly grown in the areas to which they are most suited. Sir John Russell[42] regarded this as one of the main results of the agricultural revolution, in its later phases, of Britain: from the early nineteenth century there was a development of corn- and cattle-farming in the east, cattle-farming in the west, with grazing and fattening of calves and young stock from the hill country of the west and north, or from Ireland. Similarly sheep

---

* See Thornthwaite, C. W. *Geogr. Rev.* 38, 1948, 55–94. The 'potential evapotranspiration' is the amount of water that would be lost by evaporation and transpiration from a surface completely covered with vegetation if there were sufficient water in the soil at all times for the uses of the vegetation.

were taken from the Lake District to the Vale of Eden for fattening. Intensive sheep-rearing developed on the chalk downlands of South Wiltshire and North Hampshire, since partly turned over to cropping. Market gardening had developed in areas such as the Vale of Evesham, and potato-growing in the Holland division of Lincolnshire. Russell[43] noted that 'specialization is going on rapidly, but so quietly that many people miss it . . . a pioneer tries and succeeds, and others follow; or he fails and others keep off'. Illustrating this point, an example is given that has since become well known: 'a district which is rapidly changing in appearance is Wisbech, where the old husbandry is being displaced by a fruit industry that succeeds because it accords so completely with the local conditions, and also because good transport is available.'

After the 1914–18 war, Carl Sauer[44] drew attention to the vast possibilities of economic mapping; in 1921 he said that the two main questions were the present use of the land and its possibilities. Land classification is qualitative and its desirability is partly due to natural geographical features of actual or potential fertility. But only partly, as the buyer may be influenced by the credit facilities available, the current rates of interest, the security of title, rate of taxation, characteristics of the labour supply, and access to markets. Such factors may be reflected in the land prices, or in the valuation per acre. During initial surveys attention should be given to climatic characteristics, such as the length of the growing season with its sunshine and warmth, rainfall—including evaporation conditions and variability—the snow cover of winter, destructive storms and unseasonable frosts. Equally, there should be an investigation of the soil, the slope, relief and exposure of the surface, the natural plants and animals, and the water supply, along with what Sauer calls the 'economic and social location'. No area is of necessity static: it may be changed by soil erosion, drainage or irrigation, fires, the success or failure of pest control, or improved market conditions. The work of Sauer, like that of many Americans, is partly based on the idea of the transformation of the natural into the cultural landscape—that is, the change from an area virtually uninfluenced by man to one greatly modified by his efforts. But it is interesting to compare Sauer's work with that of G. P. Marsh two generations earlier (p. 41), for each had the idea of man's transformation of the landscape in folly or wisdom, in vastly different times.

Some qualitative assessment of land is helpful. Sauer, for

example, regarded as first-class land certain areas of the United States, such as the corn- and clover-growing parts of some interior states or areas growing sugar beet, tobacco, flax or alfalfa, which had a combination of favourable circumstances. These included a reasonably long growing season, a strong and balanced soil, good drainage with freedom from floods, an adequate water supply, fields and farms of a reasonable size, no serious plant or animal pests, and suitable access to markets. A second class had land less profitable but satisfactory for crop production, ranging from wheat, barley and oats in the more favourable areas to rye, potatoes, beans and hay where conditions were poorer. This class graded into a third, or marginal type, where farmers might do well in certain years but extremely badly in others: Sauer's view was that under American conditions such land should not be recommended for agricultural development as 'we have better lands, in large amounts, available in the superior classes'. The fourth and fifth classes include two that may have a limited amount of grazing use, but the sixth is below the margin of profitable commercial use, except in places for forests, and the seventh and last is of no money value at all except possibly for building or recreation: it includes barren areas of many types due to cold, aridity, or rocks, sand, or occasional submergence. It is a valid generalization that in America many pioneer efforts have shown that land may be ruined for years, perhaps permanently; what the West European finds hard to appreciate is the 'plenty-of-land-available' idea.

In Britain, the work of the Land Utilization Survey has become widely known: every acre of land was mapped under six categories, arable land under crops or rotation grass, permanent pasture, gardens, rough pasture, forests and woods, land agriculturally un-productive through heavy building densities, mining, or industry.[45] Each category could be subdivided almost indefinitely, but in its bare essentials the scheme could be used by observers of little skill. For most of Great Britain 1:63,360 maps are available; for some areas these could not be published through war losses. There is also a series of county memoirs, some generalized maps on the scale of 1:625,000, which include maps of types of farming, classi-fication of land, and farm production: the whole enterprise is summarized in L. D. Stamp's *The Land of Britain: Its Use and Misuse*. Much of the material proved valuable in assessing the wartime possibilities of agricultural expansion in Britain, and some of it has been used as evidence in the various inquiries for planning purposes, such as the expansion of industry, housing and new town sites in such a way as to cause least disturbance to the existing economy. To a great extent, the initial aim was to see what had been done with the land of Britain and, armed with such data, to base future planning on a just assessment of current use. Apart from any immediate value, the survey gives an abiding record of the landscape of Britain as it was in the 1930's.

Such surveys need constant revision, and in recent years aerial photography has provided contemporary information on land use. From wartime use for military purposes, aerial photographs have now become valuable teaching tools in universities: after the 1939–45 war, the Channel Islands were surveyed by air in two hours and a memoir was written with this basis.[46] But it is still essential to cover the ground in detail, and at the moment of writing there are proposals for a survey of Britain with a new and more complicated classification of land use, which may be possible as the country possesses more trained geographers, thousands of whom are working in schools.[47] At least it provides a challenge. From surveys in Britain and other countries, there has now come a scheme for a World Land Use Survey, first suggested at the 1949 International Geographical Congress by S. Van Valkenburg.[48] The data will be plotted on the 1:1,000,000 maps, themselves the fruits of an earlier decision at a Congress (p. 67).

Once again, the problem of classification arises, and nine categories are suggested, as follows:

(1) *Settlements and associated non-agricultural lands.* This includes the world's cities and towns which, in highly commercial countries such as Great Britain, cover a sufficient area to appear significantly on even a 1:1,000,000 map, together with extensive surface mining areas, or areas similarly devastated by mining.

(2) *Horticulture.* This has uniformity of purpose, but diversity of expression on the surface: it may include truck-farming in America, market gardening in Britain and other European countries, large areas of gardens and allotments, and the 'garden cultivation' of tropical villages, such as those of Africa or Malaya, where the village compound may have mixed vegetables such as yams or sweet potatoes with fruit, and possibly also small numbers of palm trees, cocoa trees, bananas and the like.

(3) *Trees and other perennial crops.* Among these are plantations, including those for rubber, cocoa, coffee, palm oil, cinchona and bananas, tea-gardens, coconut groves and citrus orchards. In

middle latitudes, there are also vineyards, olive groves, and orchards for a variety of fruits, including, apples, pears, plums, cherries, peaches, apricots and figs. Cork oak groves, for example in Portugal, and perennial crops grown without rotation, such as sisal and manila hemp, come in this group.

(4) *Cropland.* A wide range of practices is recognized here. Some crops are grown on the same land year by year, such as rice, sugar cane, or even wheat and maize. In rotations, crops are interspersed on the same ground; in a separate sub-category, 'land rotation', the farms are fixed, but land is used for a few years, then allowed to revert to grass or scrub while another patch is sown.

(5) *Improved permanent pastures.* Grazing is carried on in small enclosed fields, many of them enriched by manuring: this is especially characteristic of Great Britain, Ireland and New Zealand, and some fields remain permanently as pastures but in others the grass may be allowed to grow as hay meadows. In the United States this category would include the intensively stocked dairying grasslands.

(6) *Unimproved grazing land.* Known in Britain as 'rough pasture', this covers a range of vegetation from tropical *savana* to arctic *tundra*: in the aggregate, such areas support large numbers of animals. The essential needs in the World Survey are two— first, to note whether the rough pasture is used or not; and second, to classify the vegetation types. This is by no means simple, as there may be wide differences in small areas.

(7) *Woodlands.* Here again wide differences are known, from dense to open forests and scrub, which may include the *maquis* of Europe, the *chaparral* of North America, the *mallee* and *mulga* of Australia and the acacia-thorn type of Africa and India. Swamp forests include both freshwater and mangrove types. Cut or burnt forests where natural regeneration may be expected form another sub-category. Then, too, in some forests there are forests with shifting cultivation, in some cases by wandering tribes, and a forest-crop economy where the care of agricultural holdings is combined with forestry: this is found in eastern Canada and in Finland.

(8) *Swamps and Marshes.* These are found in fresh and salt water, but are not forested. In some countries they form valuable temporary pastures in some seasons.

(9) *Unproductive land.* This category bears an obvious relation

to the poorest type listed under (6), and may include barren mountains, rocky and sandy deserts, moving sand-dunes, salt-flats and ice-fields. In some desert areas there are possibilities of irrigation, but only within restricted limits.

Study of the above list will perhaps suggest that there is little hope of the immediate production of a world map of vegetation on the 1:1,000,000 scale.[49] And it is clear that the essential aim is to provide a regional picture of land use: on pp. 127–41, reference is made to some earlier effort of a similar kind. It is not claimed that all areas will fall within one category—indeed provision is made for mixed types, such as (3) and (4) in association. But the underlying idea is an assessment of world resources, and the production of a general map may—probably must—inspire local studies that will be illuminating. The whole study of agricultural geography is rich in possibilities. To some workers, it seems that more could be done on the study of sizes of farms, and also of fields, if only because the division of the land map gives a landscape its characteristic stamp. The actual distribution of farm-houses and buildings has attracted many workers, and much delicate observation has gone into the work of various continental geographers.

Agriculture is probably still the way of life of more than half the world's population: it is certainly the occupation practised over far more than half the inhabitable area of the globe. And it is nowhere static. In the New World, the era is now passing when the transformation of the 'natural' into the 'cultural' landscape was something almost anyone might see at some time; in the Old World, there have been revolutionary changes within recent historic time. Changes of political outlook in Russia have meant the total reorganization of the agricultural landscape, and in parts of western Europe there have been enclosures, the union of small farms into larger ones, the application of manures so that yields have been at least doubled, the addition of new market-gardening areas, or the development of orchards and cash crops. Agricultural geography may easily become historical geography; those who knew the British rural landscape before, during and since the 1939–45 war will realize how great changes may be in a short period. As so many farmers are now producing goods for a market, changes in prices or the general flow of trade may cause widespread dislocation. And at all times agriculture is not merely an economic activity, but one that can be understood only in relation to physical geography.

Finally, a word must be said on the relevance of agriculture to world population. Much is heard of 'underdeveloped lands' in need of capital for economic growth to stimulate both agricultural and industrial production. The densely populated areas of western Europe and parts of North America can only support their relatively high standard of living by industry, intensive agriculture, and reliance on the purchase of goods from the world market; and as more and more people congregate in towns all over the world, so the farmer's produce must enter the market as a saleable commodity to an increasing degree. All over the world the proportion of people producing food diminishes, but in many countries the surplus available for export increases as more is available from enlarged farms, with fewer consumers in the farm households; to a varying extent the yields of crop per acre, and of milk or meat from animals, have increased through scientific advances. In his book *Our Undeveloped World*, L. D. Stamp[50] has pointed out that there are limits to the possibilities of agricultural settlement and production in the inter-tropical areas, but that the United States and other middle-latitude areas could expand their production by greater intensity of cultivation. It may be that ultimately the production of food by every nation is regarded as a general social duty; but that day has not come yet. And even if this view was accepted tomorrow, the transport of the world could not handle the produce. In many countries there are signs of a retreat from the least profitable lands, but for this very reason the evidence of pioneer agricultural settlement in the Soviet Union is of special interest. The farming frontier has oscillated backwards and forwards, and the possible maximum food-producing capacity of the world is not known. These are great human problems, underlining the thought that agriculture is itself an industry, But this is not a new idea, for in 1841 the Irish Census Commissioners[51] wrote that 'according as it receives the improvement which chemistry and physical science are daily bringing to its aid, it comes more and more clearly within that designation. The trades which minister to it, and depend upon it, will constantly increase'.

CHAPTER EIGHT

SOCIAL GEOGRAPHY

Man and environment; a time of broad views; man and the
land; urban geography; the outlook.

THE geographical use of the term 'social' has become prevalent of recent years in Britain, but with various meanings.
To some extent the word has provided an alternative to
'human' which was prevalent some thirty years ago and is still
maintained in some universities. Undoubtedly the attraction of
human geography was its social interest, its study of the way of life
of people in many places, and at many stages of civilization. Vidal
de la Blache's *Principles of Human Geography*[1] gathers up a mass of
observations made in various regional studies, and relates them to
the inherent qualities of various environments, and Jean Brunhes's
Human Geography, another classic to which readers of geography
frequently return, is an illuminating study of types of life with an
emphasis on the economic resources of varying communities as
miners, farmers, here constructive and there destructive. Much
has entered into geographical lore through these and similar books:
one finds in Brunhes, for example, some fascinating information
on transhumance and on nomadism, as well as on the life of desert
people and on oasis cultivation. A more modern book which deals
particularly with one type of environment is *Mountain Geography*,
by Roderick Peattie (1891–1955), which shows the life characteristic of many mountain people, obliged by the nature of their
homeland to evolve certain adjustments to climate, soils, isolation,
possible means of transport, in many areas under adverse circumstances.[2] It is obviously interesting to inquire why the form of a
Swiss house or village has its own distinctive appearance, and how
the seasonal migrations to Alpine pastures are arranged.

Man and Environment

Long experience has resulted in ways of living that give to
human groups their *mores*. To a great extent this way of regarding

173

people goes back to Ratzel's *Anthropogeographie*, with its American derivative volume, Ellen C. Semple's *Influences of Geographic Environment*, briefly discussed on pp. 77–81. Much of the earlier work was built around the study of environment, and the adaptation of man to environment. Many of the more primitive peoples of the world, largely untouched by outside influences, showed in their way of life a full use of the possibilities they knew, and a social code that was evolved to maintain life with some expectation of permanence. Each type of environment offered certain opportunities, varyingly used according to the quality of comprehension possessed by the human groups concerned. Out of this study of environment and people, there arose the theories of possibilism and determinism (pp. 74–82); some writers sought in racial characteristics part of the explanation of the variation in human customs from one area to another, and even from one person to another. Allowing that any study of human life in relation to the environment must be complex, one must admit that human geographers such as Vidal de la Blache never tried to solve their problems by avoiding difficulties—always they saw that man and his environment were interrelated. It became a fundamental premise that, as Hartshorne[3] has said, 'any system of division of the indivisible terrestrial unity is arbitrary, cutting through actual interrelations.'

Terrestrial unity, or the concept of the earth as one whole where all physical phenomena are interrelated, was regarded as basic to human geography by Vidal de la Blache and many more. 'The conception of the earth as a whole, whose parts are co-ordinated, where phenomena follow a definite sequence and obey general laws to which particular cases are related,' says Vidal de la Blache,[4] entered the scientific field through astronomy. He quotes E. H. Haeckel[5] (1834–1919) the German naturalist, as the originator of the term 'ecology' in 1876, and defines it as 'the correlations between all organisms living together in one and the same locality and their adaptation to their surroundings'. From the Darwinian view of the struggle for existence, or the survival of the fittest, there came the idea of migration of faunal and floral elements, and their cohesion into some form of association, never static, always interdependent, even if united in a hostile relationship. Basic therefore to the idea of man and environment is the view[6] that 'every area with a given relief, location and climate, is a composite environment where groups of elements—indigenous, ephemeral, migratory, or surviving from former ages—are concentrated, diverse but united by a common adaptation to the environment'. It was this last phrase, man and his adaptation to the environment, that appealed to many students some thirty years ago as the key to all geographical study; perhaps it is now regarded as obsolete, or even impossibly ambitious, but at least it is challenging. Vidal de la Blache[7] was fascinated by the theme of population density and said that 'the existence of a dense population . . . means, if one stops to think of it, a victory which can only be won under rare and unusual circumstances'. This dictum applies both to rural and urban communities; control of resources maintains not only vast industrial agglomerations such as those of the Ruhr valley or Central Clydeside, but agricultural concentrations such as the *huertas* of Spain or the riverine lowlands of China, which may have more than a thousand people to a square mile—Vidal de la Blache,[8] speaking of one *huerta*, says 'let us not forget that more than 300,000 inhabitants occupy an area of hardly a thousand square kilometres, all of which can be seen from the top of the tower of the cathedral at Valencia'. Long-established irrigation systems, in the Mediterranean partly of Arab origin, and in some parts of China dating back to wise rulers of many centuries ago, are still maintained for the preservation of an agricultural life of almost immemorial antiquity.

In studying geography, a kind of dichotomy between the physical and human aspects has sometimes crept in by devious means. British geography in the period before 1914 was at times and in places taught under two headings, physical and political, with the commercial aspects creeping in, occasionally with some reference to race, or to such vague aspirations as 'the earth and man'. After 1914, in Britain, human geography became a subject of study in many of the expanding geography courses, with such works as those of Jean Brunhes, Vidal de la Blache and Ellen C. Semple as fundamental texts. But all these three authors acknowledged their immense debt to F. Ratzel, whose *Anthropogeographie* was published from 1882–91: as noted on p. 78, Miss Semple's *Influences of Geographic Environment* was directly based on it. Nor was that all: deeper roots of modern human geography exist in the works of von Humboldt and Ritter, and far earlier authors whose work lies beyond the scope of this book. Having broadened the field of inquiry from the political aspects to the more generally human with stimulating effects, some geographers then fell into another habit of mind—the separation of the physical from the human aspects of the subject.[9] This arose partly in America through the use of the terms 'natural' and 'cultural' for land-scapes, based apparently on the use of 'nature' and 'culture' on the standard topographical maps. Hartshorne recognizes this 'dualism' as the current heresy and says that the study of inter-relationship is essential to geography: 'it can expect to compete for interest with the systematic sciences only if it recognizes in all its branches its own distinctive purpose, namely, to observe and analyse earth features composed of the interrelations of diverse elements with each other. While some of these features are largely independent of man and others are the product of man's work, few are either purely "natural" or purely "human".'

Consideration of the works of the classical French human geographers will show that they were deeply conscious of the unity of people and environment—of the earth and man. In this they looked back to nineteenth-century geographers, especially the Germans; equally they might have drawn the idea of interrelation-ship from the American, G. P. Marsh (pp. 41–2). They also depended very strongly on the local work of regional geographers, as study of the bibliographical references in their books shows. They were concerned to establish some general principles—not laws, like some later workers. Vidal de la Blache, for example, in an illuminating chapter on the Mediterranean lands, shows what altitudinal zones, what river basins, or riviera coasts, were most conducive to a form of settlement in which the orchard and the vineyard rather than the field was the focus of sedentary life. Yet there is no permanence in agriculture, for many areas in the up-lands have been abandoned in favour either of the towns or of lowland farms. All changes: all is in a state of flux. But there are quite clear limits to human enterprise, and there may be changes even in successful schemes evolved by the wit of man, as Brunhes[10] recognized in his statement that 'when man . . . manages to drive the North Sea back and win the Dutch polders, the risks he runs are proportional to the fruitfulness of his efforts'; this was written long before the disastrous floods of 1953 submerged some of the richest land in Holland. The more complex the social structure, the more precarious it may be, the more dependent on stable or improving economic circumstances, political security, scientific control of disease and much more. Disraeli[11] said that 'Manchester is as great a human exploit as Athens', but neither has any guarantee of permanence as study of the history of Athens will at once suggest.

In the search for clear relationships between human groups and environment, some have turned to the study of primitive peoples. This same tendency is seen in the work of social anthropologists, who turn to the simplest, most remote societies, partly for the sensible reason that unless they do so now, it may soon be too late. The techniques of relating distribution maps one to another is magnificently employed in such a work as J. G. D. Clark's *Mesolithic Age*,[12] in which for this archaeological phase the population are studied as dwellers on certain chosen types of soil, having needs and desires that could best be satisfied only in particular places. Study of the prehistoric periods leads to the concept of a steady expansion of settlement, from one type of natural environment to another, from light, easily worked soils to those less tractable but perhaps inherently more fertile. And once the value of minerals was known, further migrations followed, in search of copper, tin, lead or gold as well as iron. In turn this leads to the idea of migration, perhaps the most ancient of all human activities. The study of primitive peoples is rewarding, as the extensive literature on the subject shows; within the past hundred years there has been a rapid expansion in archaeological learning and a recording of much that is valuable in social anthroplogy. Fortunate also has been the expansion in physical anthropology which, like archaeology and social anthropology, was at one time regarded by some geographers as part of their own field.

A Time of Broad Views

The study of the social aspects of geography was more ambitiously conceived thirty or forty years ago than now. In the hands of some university teachers it was a panorama of human history from the palaeolithic age to the present day, covering the whole world, with some consideration of racial types and their supposed migrations and interminglings, all used to show the apparent geographical influences. Geography had its Huntington and history its Toynbee: though the latter was the greater of the two, both were stimulating. Many university teachers who covered—perhaps still cover—such vast fields brought to students a world view they might not otherwise acquire; they performed much the same service that is given by a general cultural course in a recently-established university which is known as 'from Plato to Nato'.

Lebon[13] has noted that Darwin's *Origin of Species* not only gave a unifying principle to the biological sciences, but revealed man as the latest product of the slow unfolding of life, maintaining relationships with the physical environment and with all other life. From this came modern 'human' geography, at times concerned with small local problems, but at others with problems so vast that comment on them, though in intention sublime, may easily become ridiculous.

Even so, a large part of the inspiration provided by some modern teachers of geography was derived from their search for general principles and an endeavour to present a synthetic view of the world. P. M. Roxby of Liverpool,[14] trained as an historian and in later years a great admirer of A. J. Toynbee's world history, splendidly stated the case for a broad human geography before Section E of the British Association in 1930; having said that the term 'human' was of recent origin, Roxby referred back to the time of Ritter and von Humboldt as one when masses of new geographical data were appearing, but in a manner 'unsystematized and unrelated' so that the subject became 'rather a torture to the memory than a stimulus to the mind'. Roxby regarded the new material available not as a preserve of the geographer but rather as part of a rich common store of knowledge brought forward for all mankind. As explorers published their observations, and as more and more data was accumulated on the nature of the physical earth, on its landforms, climate and vegetation, students of varied interests and talents could find materials challenging to their enterprise and perspicacity. Roxby held that the geographer did not claim 'a distinctive segment in the circle of knowledge *which is to destroy its very essence* (present author's italics) but a distinctive method in the handling of data common to other subjects'. One could illustrate such a premise, perhaps, by considering the contribution made by geographers to many enterprises such as town and country planning which also require the services of surveyors, engineers, architects, geologists, economists, and even sociologists and horticultural advisers. All are dealing fundamentally with one problem—the use of the land; and Roxby saw regional planning as 'essentially a conscious effort in constructive social geography, the attempt to utilize all elements in the physical environment for social well-being (as distinct from the ruthless exploitation of particular elements, e.g. coal, regardless of the wider social consequences which marked the earlier stages of the industrial revolution) and to harmonize the interests of neighbouring towns and country-sides in a common scheme in which each has its place'.

Much of this may sound Utopian to some readers, but to others it will read like the common sense that is far from common. It need hardly be emphasized that the world is changing so fast that vast social problems may arise almost before anyone has realized any dangers at all. On the one hand, in America, the spread of towns-people into the country-side with fast cars and fine roads has produced a semi-suburban belt covering many hundred square miles; and on the other, efforts in Britain, especially around London, to preserve agricultural land, partly in the form of a Green Belt, have resulted in fantastically high prices for houses and for land available for building. Study of Roxby's comments on regional planning will perhaps suggest to some that 'ruthless exploitation of particular elements' was by no means confined to 'the earlier stages of the industrial revolution': many would say that in Britain during the 1930's, or in modern America, land as one of the most permanently precious resources has been ruthlessly exploited. Roxby, however, looked to Ritter as one who saw the possibility of using knowledge as a guide to future action, when 'the world of nature as well as of morals and mind shall have been so far compassed as to make it possible for the far-seeing among men sending their glance backwards and forwards, to determine from the whole of a nation's surroundings what the course of development is to be, and to indicate in advance of history what ways it must take to retain the welfare which providence has appointed for every nation whose direction is right and whose conformity to law is constant'.

And this appreciation of the right course of action and development is not acquired merely by goodwill. Recognizing the strength of studies in economic geography. both of agriculture and industry on the one hand and commercial relations on the other, Roxby held that this must be closely related to social geography, which he regarded as primarily a study of 'the regional distribution and interrelation of different forms of social organization arising out of particular modes of life'. The term 'modes of life', widely used a generation ago and perhaps still useful, was apparently a direct translation of the *genres de vie* of the French geographers; an initial stimulus to such study came from the material made available by explorers on nomadic peoples of central Asia, on the oasis settlements of deserts, and on a variety of primitive tribes. Social organization in such communities was held to be, in Roxby's

words, 'a direct response to distinctive types of physical environment,' and many teachers found that an attractive story could be
told of 'modes of life' to illuminate the regional geography on a
climatic basis discussed on pp. 83–5. And as more and more
became known of prehistory, this too supplied examples of modes
of life having a social and economic *mores* evolved in relation,
adjustment, or adaptation—to use words used by various geographers—to the intrinsic conditions of the environment. As noted
on pp. 74–82, this led to the never-ending controversy on
determinism and possibilism which to some remains as permanently interesting as devotion to squash or tennis.

To many, adjustment appeared to be the key word in a human
study of obvious complexity. The social adaptation of the Kirghiz
or Kalmuck peoples to the natural physical conditions of their
environment might be clear and plain, but people of greater
scientific knowledge, technical achievement and educational
sophistication showed an adaptation to the environment less easy
to discern. And the revolution in agriculture and industry that
had spread gradually through the world from the mid-eighteenth
century inevitably resulted in the complete reorientation of social
organization and economic life all over the world. In the 1960's,
this is affecting to the limits of imagination the people of China as
it has affected the varied peoples of Russia: political philosophies
have a clear expression in social and economic organization. An
earlier, less dramatic, reorientation of life was seen in Denmark
under the stimulus of the co-operative movement combined with
adult education from the 1870's. Of this Roxby in 1930 spoke
with approval; but he left as an open question the future of the
peoples of tropical Africa subjected to the full shock of western
contact. Study of such peoples as those of tropical Australia,
Africa, the Malay peninsula and other parts of Asia could well
include their racial characteristics, though Roxby clearly regarded
this as anthropology and said that it was 'as necessary to find the
right relation between human geography and anthropology as
between geography and geology'. Since 1930, there has been a
great advance in anthropology, conspicuously on the social side;
but during the 1920's an interesting development was the growth
of research at Aberystwyth under H. J. Fleure on the racial
characteristics, both physical and social, of various peoples, and on
their powers of adjustment to particular climatic circumstances.

Acclimatization is a theme of considerable interest. Fleure,[15] for example, noted that 'the men of equatorial Africa, especially the true negroes, are organisms adapted to disperse heat as quickly as possible' and that though the story of skin colour was not completely unravelled, yet there appeared to be reflections of climatic circumstances: for example, 'the blackest of the black seem to live in regions with a hot, dry season ... many inhabitants of the regions of damp heat have a more chocolate tone, the dark brown modified by blood colour connected with a frequently enormous development of skin blood-vessels.' Some reference is made on pp. 80–82 to the efforts to explain much of human history by climatic circumstances: another, not unallied, line of thought was given by Fleure[16] in his suggestions (and they were clearly meant merely as suggestions) that particular racial types had particular attributes. Having distinguished the main characteristics of the Alpine, Mediterranean and Nordic elements, and noted the many cross-fertilizations between them, he observes that 'the Alpine race takes kindly to traditional village life and to industries that have grown out of it, especially such industries as have required skill in minutiae. The predominance of the areas of Alpine race in the early phases of commercial machinery is a significant modern example of the special aptitudes concerned'. Similarly broad generalizations are found in dealing with the Mediterranean type, though with a note of reservation also.[17] 'It is not that the urban culture is the special work of the Mediterranean race type, though that type has taken kindly to city life: a great culture has nearly always been the work of several types side by side. But the Mediterranean has given Europe the idea of the city rather different ... from the city of Egypt, Mesopotamia and India.' But of all the dangerous racial views, the superiority of the Nordic is the most serious, not least for its widespread acceptance in Germany during the 1930's, in some confusion with the term 'Aryan'. On Nordic characteristics, Fleure[18] is cautious, though he notes that 'the commercial imperialism of Britain in the nineteenth century drew off many Nordic elements to pioneering efforts in lands some of which are climatically rather poorly adapted to the type's special needs'.

Like many geographers of his time, Fleure was painting on a broad canvas. In T. K. Penniman's *100 Years of Anthropology* generous tribute is given to the nine-volume *Corridors of Time* survey of 'all aspects of the development from Savagery to Civilization' which Fleure wrote with Harold Peake (1867–1946).[19]

This work, eagerly read by students of geography at the time, was fundamentally of a social character: in it geographical, archaeological and anthropological ideas were interwoven with considerable effect. But in later years the volume of material available to workers in archaeology, anthropology and geography became so vast that few would now attempt so broad a picture as that given by Fleure: inevitably all these subjects have acquired separate if related identities, and diversity of theme and approach has developed within each. Penniman's book, for example, shows the vast range of anthropological inquiry, and the need for what may broadly be called 'fieldwork' not only on the physical side of the subject but in its more social aspects also. Attractive and suggestive ideas were given to young geographers thirty years ago, when some of the British academic personalities were concerned with 'the need for looking at things as wholes' and like Roxby and Fleure were prone to quote Vidal de la Blache's desire 'à ne pas morceler ce que la nature rassemble'.[20] Fleure,[21] approving strongly of General Smuts's book on *Holism*, comments that 'the whole is usually more than the sum of its parts; as a whole it has functions and relations which may not be functions or relations attached to any of its parts'. In such views it is not hard to see the influence of Darwinian evolutionary doctrines showing the interplay of all living creatures, or of social philosophies which stressed the inevitable need for human co-operation, summed up in the meaningful catch-phrases 'unity in diversity', or 'diversity in unity'.

Man and the Land

From the contemplation of the broad compelling idea, one returns to detailed study of particular problems, which probably means some local area to which careful and patient field-work is given, combined with statistical and library work. Many geographers have never been deeply concerned with ultimate objectives of a social character, and some would sincerely regard efforts to provide directives for human action as presumptuous and arrogant.[22] But sooner or later writers ask whether geography really is a human study at all, possibly because some writers on human geography appear to have studied almost everything except the people. In 1953, G. T. Trewartha[23] criticized Brunhes for implying that the covering of dwellings was a phenomenon more geographical than human beings and said that his book on 'human' geography veered off into the morphology of houses and settlements. Vidal de la Blache devoted one-third of his *Principles of Human Geography* to population, but his whole emphasis is on numerical distribution and associated density patterns. And in the case of Carl Sauer, Trewartha notes, the main concern is with the works of man. In this Sauer follows the general American practice of separating geography into two parts, physical and cultural: Sauer has said that 'man, himself not directly the object of geographical investigation, has given physical expression to the area by habitations, workshops, markets, fields, lines of communications. Cultural geography is therefore concerned with those works of man that are inscribed into the earth's surface and give to it their characteristic expression'. Trewartha apparently thinks that something has been missed even though much valuable work has been done: the neglect of population has arisen through the frequent equation of geography with landscape, and an emphasis on direct field observation of a critical and scientific character.

The equation of geography with landscape, common to many German workers is crisply stated by R. E. Dickinson[24] who says that 'geography . . . should be the study of places rather than of men'. Trewartha[25] argues that it is anthropocentric, which implies that other elements of the landscape only derive significance in relation to man or population. In recent years this approach has been seen in the widespread re-emphasis on the resource aspect of physical geography—for example in the study of the practical human aspects of geomorphology, climatology or pedology. Physical features influence human settlement in the choice of particular sites and the neglect of others, as has long been known. It was, for example, the early regional studies in Alpine areas that showed the effects of aspect on the distribution of settlements and the whole activity of the farming population; it was the natural fear of avalanches that led to the choice of certain sites for hamlets and villages for safety and also for mutual protection and co-operative action in times of disaster. Or so it is commonly believed. More recently planners have found it essential to provide maps showing areas near towns unsuited to new buildings because of steepness of slope or problems of sewerage at any reasonable cost. The study of climate and man has fascinated many people for a long time, and some of the new techniques such as those of measuring evapo-transpiration (p. 166), or of measuring current climatic oscillations (pp. 112–16) have obvious human relevance: equally the modern study of soils has its relation to crops, forestry and, in Britain, to planning if it is agreed that as far as possible the country's best lands should be preserved for agriculture.

Land use survey has a direct scientific appeal as it is based on field observations. In Britain, the writers of various county reports[26] have used this as a basis for further study including an historical review of the statistical material and of changes in land use from one time to another; a general practice was to include an analysis of some sample farms and to consider any areas or problems of special interest. But much more is involved in a study of people on the land. Traditionally many rural areas have been over-populated, or at least unable to support their families, of whom many are obliged to move to towns or to emigrate; farms may be too small to enable their holders to maintain a reasonable standard of living—for example, many schemes for establishing small-holdings in Britain during the early part of this century were unsuccessful. At present there is in Britain a widespread fear of underpopulation on the land, as even with an increasing use of machinery there may eventually be a shortage of manpower. And this fear is by no means unique, for adjustments are proceeding in many other countries.[27] Such simple assessments as the density of population per square mile in rural areas are of little value except as the prelude to further inquiry; Trewartha[28] has argued that there is a need to find the general economic density. A primitive people living from natural resources of the land, but not practising cultivation or cattle-raising, is unlikely to support more than one person to the square mile and, one may add, whenever the population rises above such a numerical density some solution must be found such as outward movements to new areas. But in more advanced societies there are social, economic and technological attributes that affect the life of a rural population, and must therefore be considered.

Of all maps showing density of population, those based on a simple arithmetical calculation of the number of people in relation to area are the most widespread and the least informative. They produce such totally misleading results as low densities of population in such areas as the western Highlands and Islands of Scotland or the west of Ireland, when in fact the limited areas actually occupied by farming families are very densely settled—indeed in the west of Ireland a problem of excessive density of settlement, or 'congestion', has been recognized to exist for more than seventy years.[29] As long ago as 1917 the Swedish geographer, Sten de Geer,* was distinguishing the areas actually settled in Sweden as a contribution to the regional study of the country. Using a dot symbol for a given number of people, this method gives what Trewartha[30] calls the 'physiological' density, by relating the rural population to the 'arable'† area or—one could say in Great Britain or Ireland—the improved land: this excludes forests, natural (rough) pasture and the like. Assessment of the agricultural density (Trewartha's term) can be roughly acquired by relating the agricultural population to the cultivated (or to use the British term 'improved' area), but not everywhere is the farming population a large proportion of the entire rural community.

Various authors in Britain have written on the distinction between the primary, secondary and tertiary elements in rural areas; the primary element consists mainly of people working directly on the land, though some authors also include any who are engaged in local mining or extractive industries, such as coal, iron ore, gravel or sand. The secondary element consists of those providing professional or commercial services to the primary group, and the tertiary element of those who have rooted themselves in the country-side but who do not provide services for either of the other groups. For this reason, they are sometimes referred to as 'adventitious': they may include retired persons, especially numerous in villages in certain favoured parts of Britain, or people working in towns but choosing to live in the country. As noted on p. 148, the spread of settlers from American cities has become a major influence in changing the land use of several hundred square miles; in Britain, due to restrictions on building and the alienation of land from agricultural use, the direct landscape changes since 1945 have been less startling. S. W. E. Vince and L. D. Stamp[31] have stated that normally in Great Britain the primary population is about half the whole in rural areas, and a

---

* Apparently Sten de Geer's first published map of population density was in an article on Gotland, see *Ymer*, 28, 1908, 240–53 'Belfolkningens fördelning på Gottland'. This article has a map on the scale of 1:300,000 with one dot to every ten people, and an inset map, 1:900,000, shows the inhabited areas by a red overprint. This method was modified (1 dot to 100 people) in the Atlas of 1917, which also had the widely-copied globes of varying sizes for towns of more than 5,000 people.

† The word arable, defined by O.E.D. as '(land) fit for tillage', is in Britain generally taken to mean land actually ploughed, though in Ireland it means the land that a farmer would plough at some time but not the fields of his farm that he retains as permanent pasture and natural meadow.

similar conclusion was reached by A. Stevens[32]; but in Ireland the present writer found that in 1946 the primary agricultural population was at least 70 per cent in most rural districts, and over 80 per cent in a substantial number.[33] These figures did not include as 'primary' paid domestic workers, nor labourers returned as 'general' rather than 'farm', though in practice many such labourers work intermittently on farms. It is possible—indeed it is virtually certain—that as the returns of income to the Irish farmers are generally less than those of Britain, the number of people engaged in secondary occupations must be fewer: certainly in Ireland the secondary population is highest in the areas having the largest and richest farms, such as counties Meath and Kildare.

Fluctuations of population have been studied in many parts of the world, and at present it is common knowledge that the world population is increasing sharply, But there is a perpetual redistribution, not least in the migration of people from rural areas to towns. In the nineteenth century, a time of great population growth, work for the increasing multitudes was found by the settlement of new overseas territories and by industrialization; but in this second half of the twentieth century the settlement of vast new areas hardly seems possible, except probably in the U.S.S.R. As long ago as 1937 C. B. Fawcett said that some of his studies had led 'to the conclusion, at present tentative, that it is not an anomaly but a fair general statement of facts at the present day, that the areas in which conditions of life are such that people can get a better living are tending to become the most crowded and that, on the converse, those areas which for any reason do not offer such prospects of good living tend to become depopulated . . . improvements in transport facilities allow that tendency fuller play'.[34] Indeed in Britain such tendencies are now a cause of comment, even of alarm, for as the maps provided by the Ministry of Housing and Local Government and the work of various writers shows, the Greater London area and the West Midlands have retained the attraction for industry that they have possessed for many decades and it is still proving difficult to attract factory proprietors to many of the older industrial and mining areas of the country. Many of these were magnets for immigrants a few generations ago, and the boom area of today may be the depression area of tomorrow.

It would be unwise, however, to accept this last statement as a general law: it is intended as a comment only. The course of economic history varies widely from one area to another, and may be influenced by discoveries of new sources of power, such as electricity or oil in place of coal or—earlier—harnessed river water. The equation of major industrial areas with coalfields no longer has the relevance of former times, and the discovery of new processes may revolutionize an industry—the example often quoted is the discovery in 1879 of the Gilchrist-Thomas process which made the ores of Lorraine valuable not only for their iron content but also for their phosphoric impurities which became valuable fertilizers. So far the story in this paragraph is economic history, but the geographer comes in with a study of the ground, of the landscape progressively altered as a result of new activities, of the changing distribution of population and the allied growth, building and, in modern times, most probably rebuilding also of the towns and industrial villages. This cannot be done without a study of the economic activities seen in mines and factories, shops, offices and warehouses, and of the communications on which they depend. All this is explanatory of population distribution, which may be regarded as primary social distribution.

Migration has long been recognized as a key to the understanding of the social aspects of geography. In recent years, some interesting work has been done by the Lund geographers on the movements of people from rural areas to towns, and of the associated problems of diminishing manpower on the land:[35] in this case, Swedish statistics have made possible work of an exactness that would be difficult to achieve elsewhere. As a phenomenon, emigration from rural areas has long been recognized; a moment's reflection will show that provided a rural population has a natural increase of population, emigration is likely unless the farming is able to absorb more workers by intensification of activity with a rising demand for manpower. Throughout much of history, 'Go west young man' (or east, north, south, according to circumstances) has been a natural method of meeting the need for emigration—in Sweden, Norway and Finland, for example, there has been an extension of settlement in the more northerly areas. But although many new farms have been made in Finland since 1945 to provide for settlers from the areas annexed by the U.S.S.R., for the present at least there is little extension of agricultural settlement in the world, but rather an intensification of farming by increasing crop yields, improving stock and intensifying horticulture, probably with a reduced labour force and more power and machinery. There are many areas of marginal farming where emigration is inevitable: within the author's observation in Ireland all the younger members of a family will leave and the parents remain till death, but no new tenant for a marginal farm will be found. Demographic study of the population structure will show an unstable situation, with a disproportionate number of old people—and ultimately a death-rate higher than the birth-rate.

Some writers deplore the existence of rural emigration and greatly regret such landscape changes as the use of mountain valleys for forests instead of small farms with rough pastures for sheep and cattle; others, however, argue that forestry may provide jobs for more people than the farming that preceded it, and that a timber industry may develop. But an even deeper social problem is the excess of population on the land in countries such as India and China, the use of much of the farm produce to feed the households, and the low cash incomes of the people, not to mention the risk of total famine through natural disasters such as flooding in China, or crop failure through unusual climatic conditions. Many older geographers will remember the impact of P. M. Roxby's articles on the population of China,[36] using the data supplied by the China Continuation Committee in 1918–19—these revealed with dramatic effect the remarkably high densities of population in the lowlands and, with other books, such as King's *Farmers of Forty Centuries*[37] led many to ask whether part of the solution lay in extending the area of settlement in places by new irrigation schemes as well as by efforts to control soil erosion. All this leads to vast questions of the meaning of overpopulation and under-population or, supposing one has ever succeeded in defining these terms, to an effort to decide what is meant by the 'optimum distribution of population'. In modern times the assessments of diets given by such bodies as the Food and Agriculture Organization are helpful, but these are generalizations which for any particular country may conceal as much as they reveal. And nothing is more difficult than assessment of poverty, especially as conceptions of a reasonable standard of living tend to rise. In 1925, Roxby[38] said that certain regions of China were 'supersaturated' with population and showed 'unmistakable symptoms of overpopulation: a standard of comfort below the average of the Orient, a basis of life so insecure that any temporary failure of crops, whether caused by drought or floods, is almost immediately followed by famine on a large scale with appalling mortality and misery as the consequence, and a constant tendency even in good years for the overflow of population'. But detailed statistical evidence is not available. If population outgrows the means of subsistence, tension is likely to arise: in 1934, the Kenya Land Commission[39] referred to 'a well-grounded and apparently unanimous opinion (held) by all the administrative and agricultural officers of the Kikuyu districts, that a state of general congestion such as will result in a depression of the standard of life is threatened within thirty years'. To what extent the disturbances in Kenya of recent years are due to such conditions is a question that one may reasonably ask.

For India excellent statistical material is available, and this was used by A. Geddes in articles which discussed the relation of population history to economic progress or regress, catastrophes such as famine and disease. Geddes[40] found that there were certain areas, such as the districts of the Punjab having excellent irrigation schemes, where the increase of population over fifty years (1881–1931) was steady and continuous. Other areas, such as the western half of Bengal, had a virtual stagnation of population, as malaria was endemic and every year the death-rate soared at the malaria season and so prevented a natural increase of population. The Deccan lands of Bombay and Hyderabad had, in the rain shadow belt east of the Ghats, years when famines were disastrous due to the failure of the millets or of the great cash crop, cotton. Such a year was 1899–1900, when mortality was high in spite of relief work; in the next great testing year, 1918, though crop failures were serious, few died from starvation partly because the government relief work was more efficiently organized. But a short time afterwards, in the winter of 1918–19, the deaths through influenza were severe, partly due to the extreme poverty of the times. Geddes's mapping of the trends of population in India focused attention of the problem areas—not least Bengal with its devastating famine in 1943; his maps have wide medical and economic relevance, and are therefore a contribution to social welfare. Here it is only possible to choose examples of special interest from a wide range of population histories, and it is claimed with some justice that 'the precise mapping of density and change of population can be of vital informative service to mankind'. A similar view was expressed by Clement Gillman (1882–1946) in his work on the Tanganyika Territory:[41] 'no more graphic way could be devised to show that peculiar problems in agricultural development, transportation, labour, and administration confront the government.' There is a clear relation between tsetse fly density and human population density, particularly where the main resource was cattle-breeding; mapping showed that more than three-fifths of Tanganyika was uninhabited. Had such maps been available earlier, Gillman believes, some of the railway branch lines might have been located differently—incidentally, probably the earliest known population density maps were prepared as evidence for the choice of railway routes in Ireland (p. 31).

In the last few pages, discussion has ranged fairly widely over many human problems, especially those concerned with the distribution of people on the land. Not for one moment is it claimed that geographical investigation, however done, can provide solutions for any human problem; what can be done is to ask the right questions and help to answer them. And at least many of the studies quoted have shown an interest in the people rather than merely in their material works. In the recent plea for a 'population' geography, Trewartha[42] asks that far more should be considered—the distribution of religion, educational qualities, occupations, and much more, including customs, habits, prejudices, loyalties; the list is obviously designed to promote discussion and thought. The range of possible mapping varies greatly from one country to another: Pierre George[43] has shown that figures for birth- and death-rates based on actual enumeration are available for only 30 per cent of the world population, and that reliable information on age and sex structure is available for only 43 per cent and in vague form for 23 per cent of the world population—for the other third, at least 800 million people, there is no data at all. Demographers have evolved efficient means of dealing with the statistical material and a great deal of social mapping has been done: it has included political sympathies in various countries, the language distributions (on these see pp. 234, 236), the proportion of adherents of different religions (p. 214), and in relation to the modern partition of India, the incidence of illiteracy, the valuation of property per head of population and many other distributions. Not all of these were originally mapped as a geographical exercise but to meet a need for apt illustration. In the 1890's Charles Booth's *Survey of London Life and Labour* included a series of maps showing the character of the inhabitants, from the criminal districts or lowest class, through very poor, mixed poverty and comfort, to fairly comfortable, well-to-do and wealthy.[44] This survey accepted the view that the house type may not be a clear indication of the nature of the inhabitants: large houses may be tenements, small houses in favoured districts expensive residences of a wealthy population, and even two sides of a street may have inhabitants of different social attributes. In the *Atlas of Finland*, published in 1925, there was an immense range of maps of social distributions, including different types of educational institutions, of co-operative societies, and even of veterinary surgeons—virtually all the conceivable distributions have been mapped (see pp. 241–2).

Urban Geography

The problems considered above are largely those of rural areas, but the distinction between town and country has become blurred in many parts of the world. American writers have shown that beyond the general built-up areas of cities there is a 'rururban' belt,[45] in which city dwellers' homes are mingled with farmlands, and a recent survey notes that 'expansion goes beyond the suburban settlements. Long ribbons of what is essentially urban development, both as regards the forms of buildings and the functions performed in them, extend far out into the rural areas along the main highways. This is a new phenomenon . . . urban ribbons . . . have no effect on settlement patterns or land uses more than a hundred feet away from the road'. But is it, in fact, so new? The author, in studying maps of British cities produced early in the nineteenth century found comparable spreads of houses along the main roads out of Manchester and London, with concentrations of housing around old villages. In Britain, the search for a suburban house close to open country has been going on for most, if not the whole, of the last two hundred years; what has changed is the extent of the outward movement through new means of communication and a rising standard of living. In other ways, too, the frontier between town and country-side becomes blurred. Reference was made on p. 156 to the sudden realization of the value of the Lorraine iron ores, and hundreds of similar examples of the use of minerals could be given: in Britain several coalfields such as South Wales, central Lancashire, Cannock Chase, the Forest of Dean, and in recent times Kent, are in areas previously having no industrial tradition. In some areas the effect of mining has merely been to implant on a farming district a number of industrial 'villages', some of which have several thousand inhabitants but few of the attractions of towns such as a wide range of shopping facilities, professional services and entertainment. The modern tendency is for townspeople to go out and conquer the country-side by building houses alien to the rural landscape, transforming old farm-houses into smart homes, and settling in villages.

From Greek and Roman times, towns have been a familiar feature of Europe, but the practice of town building is far older, as towns have for hundreds of years met certain inescapable human needs. They are fundamentally markets and service centres, long associated with the interchange of commodities, with the making and repair of goods in workshops or with agricultural industries such as milling and brewing. But they are very much more—centres of ecclesiastical and governmental administration, of education, of special legal and social facilities that may be needed only occasionally. They have had theatres, stadia and—in the case of the Romans—baths for recreation; and for hundreds of years people of the countryside have looked on towns as places of interest, to be visited at intervals. In the Mediterranean lands, and to some extent in eastern Europe—particularly in Hungary—the towns house large numbers of farmers, possibly because of the need for solidarity and mutual protection. But within the last hundred years almost every country in the world—probably every country—has seen a growth of towns and a greater proportion of the whole population in urban settlements. In 1851, for the first time in history, over half Britain's people were living in towns; but since then one country after another has seen a similar trend, and in 1960 it appears that this stage of town development has been passed in the U.S.S.R. This growth has been due to the spread of industry throughout the world and the growth of international trade; but the pulsating growth of towns, in a relatively short time, has led to a mass of social problems so great that urban study is of vast relevance. It is only possible here to indicate some lines of investigation that have been followed.

Three main lines of study by geographers have been first, site and morphology; second, economic and functional character; and third, social aspects, though the last of these has attracted a wide range of investigators who have carried the work further into the general problems of humanity. Any effort to compile a bibliography on towns is likely to prove astonishing, for though there are so many problems demanding attention on which little has been written, yet the total volume of work is impressive. It centres round the general and the particular; that is, around studies of towns on a comparative basis, such as R. E. Dickinson's *City*,

Region and Regionalism, those of A. E. Smailes and Griffith Taylor, and of French writers such as Pierre George, Georges Chabot, and Max Sorre, or around works dealing with particular towns, many of which have appeared as articles in geographical journals or monographs.[46] Among pioneer works[47] of this type Raoul Blanchard's *Grenoble* (1911) or J. B. Leighly's studies of Swedish towns may be mentioned: they have had successors, but not as many as one might wish. In Britain, a good deal of excellent material on towns has been published in the annual Handbooks issued for the meetings of the British Association[48] generally, but not invariably, in the larger cities. If one extends the view to studies of towns written by other specialists, such as economic historians, planners or sociologists, there too one finds the same general tendency to provide both general comparative and local studies; neither can exist without the other, but the real need is more local study, not only of towns as such, but of particular segments of large towns such as the various metropolitan boroughs of London. There are, however, in the most highly industrialized areas of the world urban areas so vast that they demand quite special treatment, on which some comments are made below (p. 202). Here a brief discussion is given of the particular, localized studies first, followed on pp. 199–204 by some treatment of wider issues.[49]

Individual town studies such as Blanchard's *Grenoble* show a particular concern with the qualities of the site and a treatment of a town's growth from one period to another. The vast antiquity of many French towns encourages such a treatment; especially attractive to some workers are small towns that have changed little in size for centuries and may still preserve a medieval, or even a Roman street plan. Qualities of site may be significant for thousands of years: a crossing-place of a river, the meeting of routes traversing dry ground between marshy river-beds, the convergence of ways through mountain passes, a navigable estuary with sheltered creeks for ships, a defensible position on a hill-top, all may be significant for centuries. Even in the modern sophistication of urban life, some sites will be more favoured than others and command higher prices for building land so that they are covered with larger houses and gardens. The form of a city, its urban morphology, is clearly of primary interest, as many generic studies have shown; not all, however, preserve the relics of long-past ages in their plans, long as their history may have been. In Chester there are traces of the Roman plan, modified in medieval times when the still-standing walls were built; but Edinburgh, though retaining some medieval houses in its Royal mile, owes its fine central area largely to the Georgian planning of a New Town, which was never completed though the tradition of stately building continued into the nineteenth century. The quality of life in a town is in part moulded by its site and form: unlike many other British cities, Edinburgh preserves a tradition of living close to the centre from choice among all classes of the community.

Economic study of towns at least does something to satisfy the natural curiosity on how its people live. The simplest case of all is the market town, which may be given the epithet 'town' only by courtesy and be little more than a village—as noted on p. 200, the distinction between town and village has not proved easy to draw. In many countries these towns are found a few miles, perhaps eight to ten, one from another, probably with smaller villages between them (p. 201). A dominantly rural society will depend on such places not only for trade with the outer world, but for a number of other facilities now as for centuries past. But though such towns are the basic foundation, the vast modern urban expansion has widened the range of difference between one town and another: some are dominated by industry, others by commerce and communications, others again by recreational and residential attractions, some by the trade and possibly also the associated industries of ports. In geographical and other writing, adjectives are often attached to towns somewhat casually, such as industrial, mining, commercial, holiday, residential, university, and many more, in an effort to denote quickly the main character of a place. This is no doubt useful, but it is only a beginning. The more exact work needed has taken two main forms: one of these, done by Chauncy D. Harris[50] for American cities, was to establish a 'functional' classification according to the proportion of the population engaged in particular occupations; this has frequently been done for individual towns by an analysis of the statistical material available on the occupations of the inhabitants and— by no means the same thing—the workers in factories, offices, commerce and the professions within the bounds of a town. But a more fundamental geographical technique is the mapping of the functional areas of a town, or more simply its urban land use.

Town land-use mapping has become widely practised in recent years, and particularly since the 1939–45 war through the slow but sure spread of a conviction that planning, or replanning, of towns and cities is essential to civilized living. But mapping of towns with an economic purpose is by no means new: to give one example, in 1851 a land use survey was produced for Manchester[51] (p. 230) showing the public buildings, warehouses and business premises, mills and other works, hotels with inns and public houses and private houses. The width of the streets was also shown. There may be other similar instances merely awaiting the discovery of some researcher. Many modern efforts came from American geographers who, enthralled by the potentialities of the human side of the subject, began to experiment with field-mapping in towns as well as in the countryside. The first problem was to make a classification of the land use of towns, and in doing so to strike a balance between one so simple that it is almost uninformative and one so perfect that it is too complicated to be used on a map at all. In fact the main functional areas of a town may be fairly easily distinguished. Broadly they centre around six main uses, residential, transport, public buildings, commerce, industry and recreation; also to be added is vacant sites (especially familiar to any who have worked on bombed towns) and derelict buildings. But some refinement of method is necessary according to local circumstances: in a port or a major distributing centre such as London or Manchester, a warehouse category will be needed, and in the commercial category a distinction between offices and shops could reasonably be made, with special note of such areas as the concentration of lawyers' premises so often found in large towns. Again, it is useful to distinguish banks, hotels, and possibly restaurants separately: the public-buildings category also could be divided—in one west of Ireland town the author found that it included everything from a dance hall to the convent of a permanently enclosed order of nuns! Most essential of all, perhaps, is to find some classification of houses, but this raises a number of social problems of which some are indicated below. Another difficulty is that in some cities shops, flats, offices may exist on different floors of the same building, particularly in European continental cities and to some extent in Edinburgh and Glasgow.

The need for hard foot-slogging toil has deterred many from carrying out town land-use mapping; it is, however, a most rewarding enterprise. Much has been done in Britain by the various planning authorities, but with varying thoroughness according to the quantity and skill of the labour available. Recently in Britain[52] a few pioneers have begun to map areas according to the date of building, partly to see what relics remain of early nineteenth century or earlier buildings, many of which were built as factories of a particular type, but are now changed in use though not necessarily in outward appearance. In each work architectural assistance is valuable but the art of dating within twenty years or so may be acquired fairly easily and map evidence from early surveys may be available. It is well known that many buildings in cities are pulled down and replaced by others—a process which Americans term 'sequent occupance'. In Britain, the mapping of areas of early nineteenth-century housing is a matter of social concern, as many are far below modern standards; an equally serious problem is their location, not unusually on any site left over by industry.

So far, some work along these lines shows that few areas date from one period only: for example a block of streets covering twenty or thirty acres may include some old factories in back-streets, a few rows of old houses dating back a hundred years or more, but also some bright new shops, factories, garages and perhaps workers' tenements of forty to fifty years ago on the main roads. The more accessible sites have been cleared and 'redeveloped' first, which is what might be expected. At present many of the older areas of British cities are being pulled down and altered almost beyond recognition: in Birmingham, for example, the industrial and artisan quarter round the city centre is being transformed with new roads, factories, workshops in flatted factories, and homes in flats many storeys high. In London, the East End is changing noticeably by the addition of bright new factories and tall blocks of flats in areas formerly occupied by drab rows of small houses and large grim factories or warehouses. 'Sequent occupance', which might be translated by a British geographer into the phrase 'changing land use', has an obvious economic relevance, for under conditions of free enterprise land in cities may become too precious for houses; the outward movement of population from the centres of major British cities[53] began early in the nineteenth century and was reflected in the population statistics of London by 1821, and in Birmingham, Manchester, Leeds and Liverpool by 1851. Hundreds of houses were pulled down for replacement by railways, factories, offices, public buildings and roads. But an explanation of a purely economic character is not adequate, as from the 1830's at least there was a growing concern with the living conditions of the masses in the towns, especially in times of epidemics such as cholera.

In the nineteenth-century Britain[54] the epidemics of Asiatic cholera caused many thousands of deaths: in the first epidemic, 1831–2, 22,000 died in England and Wales, 21,000 in Ireland and nearly 10,000 in Scotland. In 1848–9 the deaths were numerous and there were smaller but still very serious losses in 1853–4, 1866, 1873 and 1893. As early as 1849 it was demonstrated by Dr John Snow that the disease was waterborne and in 1855 he produced a map showing the location of deaths in an area south of Oxford Street, London; during the previous September, he had proved that one pump had contaminated water and once the handle was removed the disease almost ceased. Several other maps of cholera were produced for Leeds, Exeter and Oxford, and other towns, and in 1852 A. Petermann published a map of London showing the cholera deaths of 1832. This was based on the number of deaths in proportion to the population by the various parishes and has a scale varying from ' 1 in under 35 ' to ' 1 in over 900 '; some of the other maps actually located the deaths by streets and others by the district affected, while some—notably a series of maps of Oxford in 1856—dealt not only with the location of the disease but also with drainage (noting undrained areas and contaminated streams) and microclimatological features such as the aspect and local winds in various parts of the town. Concern about epidemics was widespread in Britain in the mid-nineteenth century and the government report on the health of towns, 1840, led to public health legislation that was one foundation of the modern administrative system of town government.[55]

Although writers on towns are legion, many problems remain unsolved, especially in the social planning of new suburbs and new towns. In Britain it has been argued that many of the first local authority housing schemes built after the 1914–18 war were deficient in social and cultural amenities, and even in essential services such as shopping areas within easy reach of the houses of the people. For this reason the conception of the neighbourhood unit swept through the offices of planners like a new broom: originally based on American studies,[56] the idea was to provide convenient shopping centres with primary schools and a variety of social facilities ranging from sporting associations to extramural lectures organized by universities, with numerous voluntary organizations. The fundamental wish was to create some form of community living, and once towns became too large for all the inhabitants to seek an outlet for their social energies in the centre, they had to find interests in some sub-centre nearer their homes. Studies of large towns have shown that many old villages transformed into suburbs had become centres of shopping, entertainment and other activities—for example London, often spoken of as a congeries of villages, has a large number of strongly-entrenched local centres—but the need was to develop such local foci of life, or sub-centres, in the many square miles of new residential areas added to town. The very fact that such local centres had acquired strength in the cities before 1914 showed that they met an essential human need. In Britain, France, the United States and many more countries, the inter-war period, 1919–39, was a time of vast expansion in urban areas: in Britain, for example, some cities virtually doubled in extent of occupied land without any great increase of population.[57] This was due to the large increase in the number of households, to the clearing of slums, to the new facilities for travel by motor-bus and private car and to a rising standard of living which gave artisans the opportunity of living in a semi-detached house several miles from their work instead of in a tiny dwelling in a narrow street near the factory or mill. What happened in Britain was that in the eighteenth century the wealthy merchant or manufacturer sought a country house in or near the countryside, in the nineteenth century the railways led the middle classes to the outer suburbs, and in the twentieth century the motor-bus opened vast new areas to the masses.

Town study has an obvious social purpose, and the geographer's work has been primarily in the mapping of distributions. Of this there are many fine examples,[58] such as those of W. William-Olsson for Stockholm, E. D. Benyon for Budapest and numerous others. Most of these studies locate the various major 'functional areas' such as offices, shops, industries with some classification of each, and some deal also with essential social distributions, such as density of population, age and type of housing and the outward spread of suburbs. Recently French geographers have made some interesting detailed studies of parts of cities, of which one somewhat unusual example is R. Clozier's *La Gare du Nord*[59] which dealt particularly with the daily flow of workers through the northern stations: the mutual interdependence of city employment and suburban residence is obviously a rewarding theme. To the present author it seems that there is scope for considerable experiment in urban studies and that the harvest already gathered should only encourage one to hope for still more abundant returns. Some workers have given particularly fruitful ideas, notably E. W. Burgess,[60] the American sociologist, who in 1923 wrote that Chicago had five main zones, concentrically arranged. These were first, the central business district (generally recognized to exist and already familiar to students as the CBD); second, a zone of transition and probably social deterioration, with some business and light industries; third, the workers' houses and factory zone; fourth, a residential area of more substantial homes; and fifth, the commuters' area within thirty to sixty minutes travelling time of the main city core. This may include towns and villages not continuously joined by housing to the main mass of the city. Burgess's work dealt also with the various national elements to be found in different areas, and he argued also that each zone tended to enlarge itself by outward expansion. Naturally heavy criticism has been levelled at this scheme, but its valuable features include a recognition of any large urban area as something constantly changing; inherent in his theory, too, is the idea of decay in towns, or 'urban blight', crisply stated by Mabel L. Walker[61] in 1938 as characterized by 'high but falling land values; congested but decreasing population; obsolete and unfit housing; a large proportion of abandoned buildings and of rental vacancies; low average rentals; generally low economic status of inhabitants; excessive crime, mortality and disease rates; high *per capita* and per acre governmental costs'. Some of these conditions will be all too familiar to students of large towns and one contribution that can be made by the geographer is the mapping and descriptive analysis of such areas, much of which in Britain has been done for individual cities by the staffs of planning offices. It is part at least of the blighted belt that Birmingham is now pulling down and rebuilding (p. 24) and that most cities are remodelling at varying rates of progress.

Valuable as all these generalizations are, it is well to remember that they are generalizations. Every town and city has some individuality, and a scheme that works for Chicago may be less satisfactory for Edinburgh. Though the cities of Glasgow and Birmingham are similar in population, they differ markedly in appearance, for in the former the main housing problem lies in the congestion of tall tenements and in the latter of back-to-back houses, squalid courtyards and small, poor-quality streets of

nineteenth-century houses. If one turns to the wider world, towns
have differences of layout, of living conditions, of means of transport, not to mention standards of living. More and more concern is
expressed in Britain at the outward spread of towns (so much that
many reviewers appear to expect every writer on urban geography
to have a policy such as advocating New Towns or the preservation of green belts or building high flats); yet in the United States
the outward move goes on with little check into an apparently
unlimited country-size. The Dutch have drawn a tight line round
their town extensions, in some cases inevitably so because of their
particular drainage problems; but the Belgians have permitted a
fusion of town and country which gives many factory workers a
plot of ground as a leisure-time occupation. In some countries,
however, artisans may not wish to spend their leisure cultivating
land. Even the post-war rebuilding of cities has taken a wide
variety of forms—flats for everyone in Poland, flats for a small
proportion of the population in Britain. In short, every town is
worth studying for itself.

Comparative study of towns has taken many forms. One might
include within this category the 'functional classification' long
followed in various ways but, as shown on pp. 194–6, this is only
a beginning; attachment of a label to any town may be misleading.
In the last thirty years the theories of W. Christaller,[62] originally
worked out in South Germany, have attracted much attention.
Working on a predominantly rural area, Christaller found that
there were seven grades of 'town' (including what other authors
have called 'sub-towns' or 'market villages') having populations
ranging from 1,000 to 500,000; he showed that the places of 1,000
were $4\frac{1}{2}$ miles apart with a service area of 18 square miles, those of
2,000 $7\frac{1}{2}$ miles apart serving 54 square miles, and of 4,000 13 miles
apart with a tributary area of 160 square miles. The large cities,
Munich, Stuttgart, Nuremberg and Frankfurt, over 100 miles
apart, were the regional capitals. In a critical study of these ideas,
R. E. Dickinson[63] has shown the need for historical perspective in
any consideration of towns and commented that Christaller 'under-
rates the importance of modern industry as an urbanizing factor'.
Many other workers have studied the areas served by towns and
villages for various services, both commercial and social in the
widest sense and have made some kind of grading from the major
cities to the villages serving comparatively small areas for immediate needs. Like many ideas, it is less modern than some

suppose, for the 1851 Census Commissioners of Great Britain[64] recognized three main categories of towns—market towns, at which men could meet weekly and return home within one day; second, county towns at which the leading citizens could congregate periodically; and third the metropolitan area. But the commissioners recognized that some county towns were growing by industrialization, that others were growing as ports, watering-places and places with mining or manufactures, or both. On the average the market towns were 10.8 miles apart, with a service area of 110 square miles. Earlier, in 1840, the Select Committee on the Health of Towns had classified towns by their main functions, as seaports, watering places, manufacturing towns, country and county towns 'with no particular manufacture' and London, the metropolis 'at all times supreme.' But the principle that each town had its service area was already accepted, and made the basis of the poor law system in 1832, when each Union acquired its workhouse in a centrally-located town or village (p. 119).

For England and Wales, A. E. Smailes[65] has divided the towns into four groups of an 'urban hierarchy': first London, at all times unique; second, the major regional capitals such as Birmingham, Cardiff, Manchester, Leeds and Liverpool; third, the larger county centres and market towns; and fourth, the simple type of market town. Each provides a certain range of services, commercial and social, but modern industrialization has made the larger provincial cities more significant than the county towns. In Britain, as in many countries, the capital is far larger than any other urban area, though one need not agree that this is an inevitable development in all countries; Mark Jefferson,[66] however, has provided a 'law of the primate city' which states that 'with few exceptions, each large nation has one primate city, much larger than any other, which best typifies the character and culture of its region or nation, and which, once having attained primacy, tends to maintain that position by attracting the greatest enterprises and talents from the whole of the supporting area.' But what of Amsterdam and the Hague in Holland, Madrid and Barcelona in Spain, Rome and Milan in Italy, or even Moscow and Leningrad in the U.S.S.R.? So far it appears to the author that it is premature to make 'general laws' or to assume that towns and cities should fall into some mathematical pattern of size and population. A more useful field of inquiry is the relation of a town to its tributary area, either by study of the towns itself, which is the method of A. E. Smailes and others, or by study of the attraction of each town to people in the country-side, as done by H. E. Bracey and various rural sociologists.[67] Each line of inquiry may usefully supplement the other; another line of inquiry closely followed by F. H. W. Green[68] has been the use of country bus services to bring people to town on market days. This last has received some official recognition in Britain by the publication of the results by the Ministry of Housing and Local Government and the calculation of the number of people in the service area of each town.

One great complication of modern town study is the growth of the industrial town and—still more—of the somewhat amorphous mining district. Both the Health of Towns writers of 1840 and the Census Commissioners of 1851 were aware of such difficulties, but their work was much simpler than that of their successors a century later. It was the growth of vast industrial districts that led Patrick Geddes in 1915 to suggest study of the conurbations,[69] or large industrial districts of Britain such as the West Midlands, the Yorkshire woollen area, industrial Lancashire and others. In 1932, C. B. Fawcett[70] extended the idea to include all towns over 50,000, and in 1951 the Census of Great Britain adopted the concept for London, the West Midlands, West Yorkshire, Merseyside, Tyneside, and Central Clydeside, and provided for these areas a wide range of statistical information. Comparable work is being done in several countries, and not least in America, where urban and suburban areas of vast extent are developing—and incidentally making the administrative definitions of towns interesting archaeological specimens. The conurbations may cover several hundred square miles, and therefore may be regarded as definite regional entities (p. 140); not everyone, however, regards conurbations as suitable subjects of study as some of them, at least, though by no means all, are expanding markedly. Study of the city-region is advocated on the grounds that to an increasing extent economic and social life centres around the regional capitals, and this has wide implications.[71] In the first place, it implies that much of the administrative system is out of date and that the apparatus of counties, states, parishes, boroughs, urban districts and the like needs revision. Second, all planning for the future must be conceived on a regional basis so that even if the present administrative units are retained with or without modification some regional authority will be formed to deal with the problems common to a whole large district of several hundred square miles or more, and integrate such planning into a national plan. Third, and perhaps most significant, study of a larger area than a conurbation may be more productive, not least because there is no permanent fixed boundary between town and country, no iron curtain between them that can remain unchanging. The countryside is permanently threatened by the garden wattlefence. To the present author, it seems that the study of the individual town, even of part of it, or of the conurbation, may be useful and not preclude the study of the 'city-region' or of the megalopolis now known to have developed in the north-east of the United States. And from past analogies, the reports of the regional planning boards of the 1920's in Britain produced most interesting results.

The Outlook

In spite of the length of this chapter, it has only been possible to indicate some of the trends in the study of the social aspects of geography. It would be possible to stress the need for more work on colonial areas, or on the underdeveloped areas of the world; fortunately the amount of published work on Africa has increased considerably in recent years, partly through the foundation of new universities with geography departments. Equally there is an advance in medical geography, partly through the encouragement given by the American Geographical Society's activities under Dr Jacques M. May.[72] Neither colonial nor medical geography could be regarded as new, for as shown on pp. 19 and 45, they have long been known: the need is to apply modern techniques to such problems. The recent advance in urban geography has been remarkable, but much more remains to be done, and then re-done, for the world's towns are changing rapidly. And this is hardly less true of the rural areas of the world: in almost every country there are changes of production, farm population, the relations of the farmers to villages and towns and the like. Any who have known some country-side intimately over a period of twenty or more years will know that changes are numerous. They may be dramatic, as in Communist countries, or they may be gradual, as in Britain and Ireland; but they are changes all the same.

Formerly the appeal of human geography, so-called, lay in its vast breadth of thought. It may seem that some modern tendencies are making it narrow and seeking enlightenment in the slum street and the untidy farm-yard. Concern with detail is often criticized but it has its purpose—the building of a new synthesis, the emergence of new and more valid and tested generalizations and above all, concern with the people. No geography can properly be regarded as 'social' unless it draws its materials from active study of men and women in their work and homes.

CHAPTER NINE

POLITICAL GEOGRAPHY

The attraction of political geography; the 1914–18 war and
after; geopolitics.

POLITICAL geography is not now the endless, meaningless
repetition of lists of towns in counties or of administrative
divisions past and present that caused a vivid revolt against
such teaching fifty years ago, but rather a fascinating and challenging study of states, nations, their resources, their relations one to
another, all based on their regional geography. No longer is there
a struggle between political and regional geography; each is seen
to be contributory to the other, largely because at the end of the
1914–18 war a number of continental workers made illuminating
geographical studies of the countries newly established in Europe,
notably J. Cvijić for Yugoslavia and the great Frenchman, E. de
Martonne for the whole of central Europe, including Germany.
And of Isaiah Bowman, Director of the American Geographical
Society from 1915–35, A. G. Ogilvie† (1887–1954) wrote[1] (in 1950)
'it is not yet possible to assess Bowman's influence, direct and
indirect, upon international affairs, during the past thirty years. It
is certain, however, that no man has applied so consistently and
effectively the discipline of geography to difficult political and
economic problems in war and uneasy peace'. To students Bowman is known at least by his book *The New World*, first published
in 1921 and therefore now belonging to a past epoch of twentieth-
century history yet still abundantly worth reading though supplemented, rather than supplanted, by works of a later generation.
Having served as Chief Territorial Specialist on the American
Commission to negotiate Peace in 1918 and 1919, Bowman was
even more closely connected with high policy from 1940 and was
one of the architects of the United Nations. The objectivity of
Bowman's work is one of its most remarkable qualities for, as
shown on pp. 215–23, some distinguished continental geographers
were far less detached, not unnaturally because they had a personal
stake in changes on the political map; de Martonne's work was

exceptional in showing an evenness of temper and shrewdness of approach not common to all French geographers. In Britain, H. J. Mackinder stands out as a stimulating writer on world issues rather than a writer in any detail on political geography. Bowman showed the wood and the trees; Mackinder the wood only; Cvijić his own well-studied trees.

All of these men would have been famous geographers without their writings on the political aspects: Cvijić, for example, is chiefly known for his work on limestone landscapes both in Yugoslavia and in a wider generic context. Indeed he spoke of himself as primarily a physical geographer[2] who, during the Balkan wars of 1912–15, was forced to consider questions of political and human geography. And he was not ashamed to admit an interest in people: every year from 1887 to 1915 he had wandered through the countrysides of what later became Yugoslavia, studied the peasants and talked with the educated people he found. Consequently *La Péninsule Balkanique* includes a great deal of material that would now be regarded as sociological rather than geographical, and like some writers of a later age he would now be reproved by reviewers for spreading his net too wide. In fact, the position of the political geographer is in many ways peculiarly difficult. Some writers on the subject have been attracted by racial arguments, drawn from anthropological sources marked rather by the presentation of hypotheses than by final conclusions; others again have shown an agnosticism, or even atheism, that makes them impatient of such matters as the distribution of religious allegiance, a vital factor in the Balkans or in Ireland; others again are Marxist, even Communist, with judgements coloured accordingly. But of all the categories the geopoliticians are the most suspect, for though their arguments on the right arrangement of political units apparently rest on a geographical foundation, their aims are fundamentally political rather than geographical, with the result that they are not in favour with the vast majority of geographers. Nor are they favoured by political scientists, some of whom would cheerfully hand them over to the geographers. They are discussed briefly on pp. 224–5.

It was the rearrangment of the world following the 1914–18 war that gave geography, especially in Britain and America, much of the impetus of the following forty years. But the roots of modern political geography go deep into the past, and in such permanent European problems as the definition of the German lands one is compelled to consider such ancient stories as the eastward movement (the *Drang nach Osten*) of Germans into Bohemia, the Danube lands, the former Baltic states of Lithuania, Latvia and Esthonia and indeed the whole of eastern Europe. Cvijić begins his work by saying that there is '*une répugnance évidente*' at the description of the Balkans as 'Turkey in Europe': it had also been called the Byzantine and the Greek peninsula. Inevitably history and geography meet in this field, and E. A. Freeman's *Historical Geography of Europe* which first appeared in 1881 described its aim as 'to trace out the extent of territory which the different states and nations of Europe and the neighbouring lands have held at different times in the world's history, to mark the different boundaries which the same country has had, and the different meanings in which the same name has been used'. And in the nineteenth century spread beyond Europe, the need for geographical study of the new colonies was clearly appreciated though, as noted on p. 52 the opportunity was grasped far more efficiently in France than in Britain, where work on a comprehensive scale came only after the 1939–45 war, and then largely through the enterprise of the staffs of new universities in the Commonwealth. In Germany there was a notable growth of colonial geography from 1936, very clearly revealed at the International Congress of 1938 and obviously related to the expectation that some at least of the colonies removed from Germany in 1919 would be returned. At that time, too, German geographers were penetrating many fields of political tension, conspicuously Spain. Much of the modern work on British dependent territories can hardly be called political, but rather economic and social, as it consists of applying general geographical techniques to areas where they have not previously been used. Similarly one may group some of the notable work on world food supply as chiefly economic in aim.

The Attraction of Political Geography

Political geography has the obvious attraction of presenting a broad view of affairs, and its generalizations can at least give rise to stimulating debate. Indeed, Hartshorne,[6] having spoken of Mackinder's view of world strategy, notes that his purpose was chiefly 'political and practical' and that his theory of the heartland was far less firmly grounded than his analysis of Britain's sea-power in *Britain and the British Seas*. World power groupings inevitably concern all thinking people; the defeat of Japan in 1945 has made the 'Greater Asia Co-prosperity Sphere' a matter of mainly historical interest, but the modern reorganization of China in agriculture and industry and its potential military strength in Asia, especially in relation to India, raises obvious problems. And the technical revolution in Russia, both in agriculture and industry, combined with its rapid growth of population and notable town expansion, compels attention—indeed, it has led many people to think that Mackinder was at least partially right after all. Even for mere definition the geographer may be needed: early in the 1939–45 war the present author was asked by a barrister to provide a definition of the 'Western Hemisphere' for a case in the courts, and fortunately the *Geographical Review* provided an answer which could be produced as evidence.[7]

But the perils of generalization are endless. The safest approach to political geography is through regional study and for this reason many who wished to know more of the political aspects of South America would turn to such a work as Preston James's *Latin America*[8] or for inter-war Europe to standard works such as the *Géographie Universelle* volumes as well as to J. Ancel's *Manuel Géographique de Politique.Européenne* (1936 and 1940) or H. G. Wanklyn's *Eastern Marchlands of Europe* (1941), a most stimulating study of the countries existing in 1938 between Germany and Russia, or to the same author's material on Czechoslovakia made after the forced migration of the Sudeten Germans, following the 1939–45 war.[9] Nor would one turn only to such works, but to historical sources, economic surveys and the like. What many seek is a guide to action, and the political geographers like Cvijić who worked nearly fifty years ago obviously had their eyes fixed on the makers of peace treaties. Historical arguments could always be used, and economic and strategic considerations urged, but no evidence was more vital than the distribution map, showing the percentage of people speaking various languages, belonging to different religions or voting for certain political parties, as well as the maps of communications, towns and their trading areas, with many more. To give Poland access to the sea appeared essential, yet Danzig at the mouth of its great river, the Vistula, was undeniably a German town; Yugoslavia needed ports on the Adriatic but the allocation of Fiume and Trieste raised immense difficulties. Even in Ireland, the retention of six counties as part of Great Britain cut off Londonderry from its main trading hinterland and put a customs frontier at the end of a main street in Clones. Railways and roads were crossed by new national frontiers of powers that might become hostile: for example, after the Munich settlement of the Czechoslovak boundary in 1938, some railways were sliced by enclaves of German territory. And during the inter-war period the Poles built a special strategic railway from Katowice to Gdynia (completed in 1933).

Frontiers may be of little account as features of human geography or of great significance. Reference was made on p. 69 to the apparent wisdom disproved by experience of making a frontier along the crest of the Andes—which crest and which water-parting? A river boundary may be vitiated by changes in the course of the stream: one example is the Foyle, which for part of its course forms the boundary between the Republic of Ireland and the United Kingdom—the boundary remains when the river moves, with the result that for more than a hundred years there have been disputes and even lawsuits on fishing rights. Such an example may be trivial, yet it is part of a much larger problem: when Ireland was partitioned in 1922, six counties, with their unaltered boundaries, remained in the United Kingdom and these ancient boundaries received the status of an international frontier that has proved difficult to guard and patrol in times of political tension—smuggling goes on perpetually. It was not intended that this frontier should be permanent, and a commission sat to study its rectification, but no report was issued, however, and in 1925 the matter was permanently shelved in return for financial concessions to the Dublin Government. In the material produced for general circulation, essentially geographical questions are raised such as the market areas of the country towns (apparently from local observation), the distribution of religions which was assumed to be a measure of political views (and with exceptions no doubt was), the hinterland of various ports and the network of railways; but little was said of roads, naturally enough in the earlier phase of motor transport. Historical and economic factors were also discussed, and it was pointed out that Article 12 of the Anglo-Irish Treaty said that the revised boundary was to be drawn 'in accordance with the wishes of the inhabitants, so far as may be compatible with economic and geographic conditions'. Virtually the same formula was used in the Treaty of Versailles[10] for the areas in which plebiscites were held—Upper Silesia, Allenstein and Marienwerder, disputed between Germany and Poland; Schleswig, disputed between Germany and Denmark; and Klagenfurt, where a 1920 plebiscite favoured union with Austria rather than Yugoslavia. One need not stress that boundaries remain contentious: as early as 1921 it was agreed at a meeting of German geographers in Leipzig that all the lost territories must be indicated on maps.

Preoccupation with states in their international relations marks political geography; but this is only part of the story, for it is also necessary to study states as entities and to analyse their territory and peoples—as noted on p. 215, this has led to consideration of the 'core' or 'nuclear' areas of countries.[11] D. Whittlesey stresses the 'core' idea strongly and says that 'in nearly all states the nuclear core is also the most populous part'. Such a core may be a physical unit, for example the Bohemian basin, or an area inhabited by a group of people conscious of nationality such as the Czechs, who unfortunately occupied only part of the Bohemian basin. In some countries of the 1919–39 period more than one group was combined: for example the Czechs, Slovaks and Ruthenes, or the various groups of Yugoslavia. And in both these countries there were substantial minorities potentially hostile to the government. In both Czechoslovakia and Yugoslavia differences of history, outlook and—to some extent—religion between the major groups made union difficult, but over and above these problems the substantial minority groups were a further source of trouble. The Versailles Treaty had special provisions for such groups but the temporary downfall of Czechoslovakia in 1938–39 was due to externally-fostered agitation among the German and Hungarian communities as well as to efforts to separate the Slovaks from the Czechs, and the attack on Poland which opened the war was heralded by German propaganda on the deplorable fate of their people governed by Slavs of Warsaw. Voluntary migration and forced movements of people since 1945 in Europe have imposed untold suffering on millions and altered the distribution of population and of states: Poland's boundaries are now pushed west into what has for centuries been regarded as German soil but the substantial Russian minorities of the 1921–39 Poland have been incorporated within the U.S.S.R.

Within each state there are many political distributions of supreme interest. Roxby[12] spoke of regional loyalties, such as 'Kentish and East Anglian patriotism' which 'without entirely disappearing, were gradually merged into the larger stream of English patriotism'. Fears have often been expressed of the standardization of man by easy communications, mass entertainment, centralization of government and the nationalization of industries and services: it is not for nothing that people should wish to keep a regional loyalty to Bavaria, to Brittany or even to Cheshire. In the first phase of post-Versailles thinking, some intellectuals regarded themselves as 'international', in some cases building this on a strong national loyalty, but social thinkers stressed that some form of local attachment was clearly necessary for corporate living, and the word 'regionalism' acquired the sense of allegiance to a neighbourhood of some kind, such as a town, village, district, county, province. Presumably we all need some focus of living, perhaps the 'neighbourhood unit' of the modern planner; it is not without significance that this term came from New York, but it is hard to define, though one cynic spoke of it as 'the area in which you are gossiped about'. One cannot assume homogeneity of outlook in any such political division as a state: Hartshorne,[13] for example, quotes a pioneer study of 1915 which showed that although the state of Tennessee appeared to be a homogeneous unit on a political map, faithful to one party, in fact over a third of its counties had majorities of the opposing party, and the districts revealed by mapping these data showed marked correspondence to the economic and racial differences in the several portions of the state—the Mississippi lowlands, the Nashville Basin, the Tennessee valley and the mountain region.

Far from needing less political geography, we need more. As the study of government advances in the universities, so the geographical distributions which are among its concerns must receive attention. To return to Ireland, one could not possibly study its politics without some conception of the basic religious differences combined so subtly with social attitudes that—in the absence so far of adequate sociological investigation—seem indefinable though they provide rich material for journalists and novelists. Nor could any study of its economic geography be successful without a recognition of the political division of the country into two units differing widely on policy and only now, after forty years of existence, beginning to show a co-operative spirit. On a more local scale, much interest has been aroused in various countries by the mapping of election results: in Holland, for example, this is strongly influenced by the distribution of Roman Catholics. In Britain this factor hardly enters into elections at all, for the nineteenth-century phase of Nonconformists voting Liberal and Anglicans Conservative has gone—if indeed it ever existed. Of the two major parties, the Conservatives have their largest majorities in wealthy residential districts and the Labour vote is highest in mining areas, but the division is by no means according to income-level only, as in the north of England and in Birmingham there is a considerable artisan Conservative vote (once Liberal) and the Labour party has some support in all classes of society. But in the main the rural vote is Conservative, though Liberals have seats in fringe areas of Wales and Scotland, as well as a few others. Local government elections unfortunately attract too small a proportion of the voters to be indicative of sentiment, yet one would be surprised to find a Conservative borough council in the Rhondda valleys or a Labour one in Harrogate. After every election, the author looks at the maps in the papers and wonders what exactly they indicate, especially as the final results appear to be swayed by a small fraction of the voters who 'float' from one allegiance to another. But such social distributions are worth studying: one would, for example, be intrigued to see in what parts of France Communism is most entrenched and to relate this to the comparative hold of the Church on the population. But these are social distributions that can be mapped more plentifully than they have been so far, possibly with interesting results.

The 1914–18 War and After

The most vital result of the 1914–18 war was the break-up of the Austro-Hungarian dual monarchy and the exclusion of Turkey from all save one corner of Europe. Contrary to expectations, German militarism was not subdued for ever nor was Russia permanently relegated to an interior-continental enclosure from which it could hardly be expected to go forth into the world power arena. In the Pacific ocean Japan was immensely strengthened by strategic gains from Germany, removed as a colonial power from oceanic islands; China remained torn by civil wars and revolution. For the first part of the inter-war period it seemed likely that the League of Nations could keep the peace but such hopes gradually diminished in the 1930's. Many who were students at this time will remember the attraction of the basic geographical idea of the unity of the world, developed not only from those regional writers who had taken a cosmic view but also from Vidal de la Blache and others whose writings stressed the unity of humanity. But as the world economic crisis developed, tension grew not only between states but also within states: apart from such obvious manifestations as the Nazi movement of Germany, and the increasing militarism of Japan, the newer countries of Europe such as Poland, Czechoslovakia and Yugoslavia had to meet in varying degrees severe economic problems that inevitably hampered both their work of economic consolidation following the war and their social task of welding various groups having different sympathies and histories into national units. It is an over-simplification of the Versailles treaty to say that it created national units, but it came nearer to this ideal than any previous European treaty and its catchword 'self-determination' was by no means disregarded.

Nationality defies definition as a world phenomenon, except perhaps as some consciousness of unity among people, some desire to belong to a community having corporate feeling but not necessarily identity of views. The American scholar, Leon Dominian†[14] (1880–1935) said that 'Europe was stirred to the consciousness of nationality by the French Revolution. Nations began finding themselves when the doctrine of man's equality, proclaimed on French soil, found responsive welcome among the peoples of the world'. This may be so, though one must add that some form of national consciousness is far older than the nineteenth century— who could deny that it is deeply ingrained in the Bible? Dominian[15] regarded the decision made at the treaty of Paris (1814) to unite the German States into a single confederation as the beginning of German nationality: for the first time in history thirty million Germans were effectively welded into one potentially great power. French writers have stressed the deep historical roots of French nationality[16] and regarded that of Germany as something slowly induced during the nineteenth century by the statecraft of Prussia, combined with industrialization, military and naval power, and the growth of Berlin as a centralizing capital. Franz Schrader† (1844–1924), famed as a cartographer, gave a notable expression of the French view of Europe after the 1914–18 war.

Although the French revolutionary ideas gradually permeated every part of Europe, they were not the basis of the Treaty of Vienna in 1815, in which, it has been said, Europe was treated as if it were a blank map which could be divided into arbitrary districts of so many square miles or so many inhabitants, without reference to nationality or the wishes of the people.[17] France had a frontier on the Rhine which gave her Alsace and Lorraine, retained until 1871; Poland remained divided between Prussia, Russia and Austria; the Austrian Empire, transformed in 1867 into the Dual

Monarchy of Austria–Hungary, kept its control with varying success over Slavs it never managed to absorb, and the Turks gradually lost power in the Balkan peninsula only to be ejected from almost the whole of it in the twentieth century. No attempt was made in 1815 to give recognition to various nations and of all blunders the greatest was to unite Belgium and Holland, separated again after a war in 1830. It may be that the Pan-Slav movement was a foundation cause of the 1914–18 war—at least the initial shots were fired between Austria and Serbia—but the real fears of the west lay in German dominance of Austria–Hungary, as part of a *Drang nach Osten* policy which might bring Germany into Asia Minor, Mesopotamia and finally India. Certainly Dominian in 1917, arguing mainly but not exclusively on linguistic grounds, said[18] that 'the Near Eastern Question cannot be settled without cutting away from Austria-Hungary and uniting with Serbia and Montenegro, all the southern Slav provinces of the Hapsburg crown . . . Southern Slav unity and independence are both neces-sary to Europe. Serbia, or rather Serbo-Croatia, or "Jugoslavia", is reared on a land-gap that provides Europe with a gateway to the east. The freedom of Balkan peoples and to a great extent the freedom of Europe depend on the power of the southern Slavs to hold the gate.'

Concern with the distribution of language was general during and following the 1914–18 war. Dominian[19] spoke of his work as 'a study in applied geography' and of his aim as the tracing of the connection between linguistic areas in Europe and the subdivision of the continent into nations; he further says that 'language exerts a strong formative influence on nationality', subject to other in-fluences, and in many cases, linguistic frontiers deserve recognition as 'the symbol of the divide between distinct sets of economic and social conditions'. This may well be more apparent in eastern Europe than farther west, particularly in Turkey, where Dominian spent his early years. In the industrial districts of Upper Silesia, so intricately mixed, the Germans were more abundantly repre-sented in the wealthier classes than the Poles. Other difficulties arose where Germans were predominant in the towns and Poles in the countryside, as for example in Poznan (Posen), though in such areas of conflict there are generally allegations that each group is trying to buy land to secure stability in both town and country-side; the continued resistance of the Poles in Poznan is generally ascribed to religious differences as the Catholic Poles are easily distinguished from the German Lutherans. And the Poles were no less easily distinguished from the Orthodox peoples of Russian stock who were incorporated in Poland in 1920. Perhaps with some exaggeration, Dominian[20] commented that 'to a notable degree areas of homogeneous language in Europe have been spared the havoc of battle or siege . . . linguistic borderlands have always been scenes of armed struggle and destruction'.

Historical justification for the existence of states is sometimes regarded as depending on a 'core' or 'nuclear' area (or areas), where each state first achieved some identity or individuality. Possibly in the New World such areas would be the sites of the first settlements, some of which have become major cities and foci of highly significant areas, such as Sydney, Melbourne, Buenos Aires, or Rio de Janeiro. A. G. Ogilvie[21] has given a list of such core areas for European nations which is interesting if controversial: in Norway the core areas were the successive capitals of Trondheim, Bergen and Oslo, and in Sweden, Svaeland, then extended westwards through the conquest of Gotland by the fifth century; in Belgium the historic towns provided nuclei and in the Netherlands, the successful resistance of Holland and Zealand to Spain in the sixteenth century provided a 'core' of sentiment rather than of area for a growing nation. Spain, in so far as it achieved unity at any time, achieved it by its crusades against the Moors from bases at first in Asturias and Aragon; but Portugal became separated in the eleventh century. Switzerland developed around the original three cantons with Luzern from 1332. But perhaps the most fascinating of all is Russia, having Novgorod and Kiev as Norse-influenced centres in the ninth century, and growing round the kingdom of Muscovy from the twelfth century, spreading across vast plains at times reaching one sea and at times another and finding, for much of its history, its vastness an embarrassment.

Self-determination as a principle of political geography after the 1914–18 war gave Europe a number of boundaries that seemed highly unnatural. Czechoslovakia was based on the whole of Bohemia to its natural frontiers at the price of including some three million Germans; its Slovak section was carried from the mountains to the Danube to include a large and resentful Magyar group, and Ruthenia, south of the Carpathians, was pushed in presumably because the Poles had the mountain crest as their only 'natural' frontier. And if one carries the story further, study of the political map of 1938 shows that the middle Danube basin, apparently designed by nature as a regional entity, was shared out by Czechoslovakia, Austria, Hungary, Yugoslavia and Romania— even the U.S.S.R. now has a foothold here since it acquired Ruthenia in 1946. Allowing that the defeated powers, Austria and Hungary, stood to lose, the gains of others were considerable on linguistic evidence and strategic argument. Though apparently regarding the break-up of Austria–Hungary as inevitable, Dominian[22] made the forecast that Germany would try to pick up any remnant of Austria that remained, which happened in the Anchluss of 1938. Dominian[23] speaks also of the Rhine valley as 'a natural region . . . an area of German speech' and frankly admits that 'the startling preference of Alsatians for French nationality cannot therefore be substantiated by geographical evidence', but indicated 'the persistent influence of the human will . . . . swayed by feelings of justice and moral affinity rather than by natural considerations'. Alsace and Lorraine were treated in detail in Vidal de la Blache's last book *La France de l'Est*,[24] which was primarily a regional study though he urged also that it was essential that France should have access to the Rhine and the control of Strasbourg. Plebiscites sound to the uninitiated the acme of fairness, yet recriminations inevitably follow even in such cases as Schleswig where, it was alleged by Germans, neutral Denmark was unfairly favoured. And in the case of the mixed Polish–German population of the Upper Silesian industrial area, though both the Poles and the Germans claimed that the division of so highly integrated an economic unit was impossible, a division was nevertheless made by arbiters of the League of Nations.[25]

Cvijić's arguments for the creation of a Yugoslav state were based on a thorough survey of the country which begins with physical geography and ends with what would now be called sociology. In 1888, at the age of twenty-three, Cvijić[26] began to publish papers on the *karst* and in his thirties he worked on the geology and tectonics of the Balkans, for which he produced in 1900 the first geological map of the central parts of the peninsula: this led to an interest in the character of valleys and in limnology, on which papers appeared up to 1914. But he was no dehumanized geomorphologist, and in the preface to his *La Péninsule balkanique, géographie humaine*,[27] published in French in 1918, he says that he diverges from the great masters of human geography, Ratzel and Brunhes, as they excluded man from their work to an unjustifiable extent. In precept and in practice he was impatient of limitations in dealing with people and over half his book is given to the *Caractères psychiques des Yugoslaves*, which he found a 'simple and direct study' as the sixteen million Yugoslavs from Klagenfurt and Laibach on the north to Salonika on the south and from the Adriatic to the Black Sea (that is including the Bulgars) had not been 'made uniform' by civilization.[28] But he was deeply aware of the effects of past invasions among a people so dangerously placed[29] between Turks, Magyars and Germans (Austrians), though as it turned out it was the Italian–German axis that proved the threat twenty years later. Within the south Slav territories there had been many peaceful migrations, for example of upland people (areas from 1,000–1,500 m.) to the rich lake platforms of Sumandija; transhumance[30] survived also and was most carefully studied. Cvijić includes a survey of the historical influences that affected the peninsula and an interesting map of climatic types; he gradually works up to a study of natural regions vital to his later argument. Also given is a map of *zones de civilisation*,[31] of which the four main types are patriarchal, modified Byzantine (or Balkan), Mediterranean and Italian, and Turkish, together with areas showing central European influence, western European (towns only) and the areas having a national civilization showing both eastern and central European influences.

Like the French geographers of his time, Cvijić deals thoroughly with village and house types, all of which are illustrated and mapped; also typical of the time was an interest in racial types and languages, but the main purpose was to give an impression of the character of the various groups within the country. The people of Sumandija[32] (in effect the Morava valley and what was Serbia in 1912) were regarded as having a strong consciousness of nationality, a good and healthy if somewhat formless democracy, great moral and spiritual bravery, a talent for initiative and learning, and a capacity for formulating ideas and carrying them out. But the Dinaric Slavs, though sharing many admirable qualities of moral and spiritual idealism with the Serbs, showed such primitive elements as animistic elements of religion, with a belief in spirits of the water, of the ground, of trees and of fairies. Here, too, there was extreme pride of ancestry as in Montenegro, where the ancestors might be known for fifteen generations, and a tribal spirit with deep feuds still existed though it was in decline. Cultivation was limited to small areas surrounded by *karst*.

Although Cvijić included the people of Bulgaria as south Slavs, he regarded them as a definite eastern group whose main modern route was the Sofia–Adrianople railway and whose western boundary was the mountains beside the Vardar valley, the Rila, Rhodopes and Ossogov.[33] The new Slav state must be built around the Morava-Vardar valley, as all the people of this great route had the same type of life and agriculture, but Salonika was not claimed as it was 'non-Slav': on the north the frontier could possibly follow the Sava and the Danube to the Iron Gate, though on the north side of these rivers, in the Pannonian Basin, Croatia, Slavonia, Syrmie, Backa and Banat, the Serbo-Croats were in a majority or at least 'an important part of the population'.[34] Cvijić speaks with warm admiration of Serbs who had settled north of the Danube and preserved their own language and Greek orthodox faith; he regarded the Croats as under Catholic and clerical (fatal word— so often used) influence, and therefore inherently opposed to the Orthodox Serbs.[35] There was apparently an idea of a Croat-Slovene state distinct from Hungary and Austria, but many Croats were attracted to the idea of Yugoslav unity. Cvijić did not claim to be impartial[36] and his view was that 'there is no economic or intellectual independence without political independence'. His views did not meet with universal acceptance and in 1919 Commander Roncagli,[37] of the Italian Navy, wrote an article in the *Geographical Journal* claiming that Cvijić's recent writings attempted to demonstrate Serbia's ethnological predominance in the northwestern part of the Balkan peninsula, as far as the Adriatic sea and northern Atlantic. Apparently the argument rested on the geographical and geological facts that the karstlands and the Dinaric mountains were integral parts of the Balkans and Italy should find her limits in the middle of the Adriatic sea and not on its eastern shores. And Cvijić proposed to use the river Isonzo (west of Trieste) as a boundary and therefore to include in the so-called Yugoslav territory 'that part of the province of Udine in which Slav populations have been living for centuries and are now completely Italianized'. The Commander argued that the Dalmatian anchorages and harbours were essential to Italian security. And on a somewhat higher academic plane, O. Marinelli (1874–1926) of Florence,[38] argued that Istria, including the Quarnero islands (Veglia, Cherso and Sussin) 'resembles a typically Italian region both in its physical features and in the human occupation of its soil, especially its arboriculture'. Even more markedly, he added, its cities were Italian. He shows the difficulties of Trieste and Fiume, which had developed as Austrian and Hungarian ports respectively but possessed a strong allegiance to Italy, especially in the case of Trieste.

Hungary's losses after the 1914–18 war at the Treaty of Trianon 1920 by cessions to Czechoslovakia, Romania and Yugoslavia were regarded as catastrophic and few would envy her geographer Prime Minister (1920–1 and 1939–41) Count Paul Teleki† (1879–1941). In 1919 under the editorship of L. Lócky (1849–1920), President of the Hungarian Geographical Society, a geographical, economic and social survey was published 'on the eve of peace negotiations':[39] the material was extracted from an unpublished survey of the Hungarian part of Austria–Hungary prepared before the war. The survey as a whole is academic rather than polemic in tone but it is clear that the authors had no conception of what was coming. They spoke of the northern natural [sic] frontiers as 'sharply defined' by the highest line of the Carpathians from Bratislava (Pozsony) to Orsava on the Danube—that is, including the whole of Slovakia and Transylvania (the latter was given to Romania). They found possible southern boundaries more difficult to draw and admitted that the Drava and Sava rivers, continued into the lower Danube, had long demarcated the political frontiers separating Hungary from Slav territories. Nevertheless, the authors urge gently that in the Hungarian lowland and the broad mountain valleys opening from them 'we find everywhere similar habits of life', and they state that similar characteristics were to be found among the peoples around the Drava and the Sava. Incidentally, the authors recognized what is often forgotten —that in a mountain area the peoples of valleys and lowlands may have more association with the surrounding lowlands than with the scenically dominant hills. 'Nowhere in the world,' say the authors, 'does there exist so uniform a closed basin as that flanking the central Danube . . . identical natural conditions . . . compel the peoples living in it to friendship and mutual agreement.' But this proved a pious aspiration. The survey points out that there had been considerable drainage works, says something of the growth of industry, of forestry and of mining, and deals also with history and education.

Many forgotten papers dealt with the frontiers and states of Europe after the 1914–18 war:[40] in the *Geographical Journal*, for example, Miss M. A. Czaplicka[41] gave a most careful survey of

Poland and reached the conclusion that ethnographical and economic Poland can be defined by a figure formed by lines drawn from
Vilna to Lwow, Lwow to Cieszyn (Teschen), Cieszyn to Poznan
(Posen) and Poznan to Gdansk (Danzig). In the west her line
corresponds closely to that of Poland in 1938 but in the east it lay
some thirty to fifty miles on the Russian side of the 'Curzon line'
suggested at the peace conference and reproduced with modifications in the present eastern Polish boundary.[42] E. Romer† (1871–
1954) the Polish geographer, wrote a number of papers advocating
a new state with widely-drawn boundaries;[43] in one of these he
said that in 1916 the Poles had taken a census which showed that
'the importance of the Polish element in Lithuania is incomparably
greater than had been supposed . . . the Poles are the only element
there who possess real political and constructive qualifications'.
He added that in the area around Grodno and Vilna there was a
mixed population of Poles, White Russians, Lithuanians and Jews
but though no group was in an absolute majority the Poles were
the most numerous. In another article, Romer argued that Galicia
was Polish: in the west 96 per cent of the population was Polish
but in the east the Ruthenians were 'only 59 per cent of the
population' *[sic]*. He argued that the Poles had 'superior social
energy', were better farmers, more thrifty by temperament and
paid three-quarters of the taxes. They were over twice as
numerous as the Ruthenians in trade, industry and the liberal
professions, and were strongly represented in the oilfields, though
on this Romer was rather reserved as most of the capital was
foreign. Enough has been said to make it clear that geographers
were active in propaganda during and after the 1914–18 war.
Indeed, in 1919 de Martonne wrote in Paris of a tour of Bessarabia,
in a light-hearted travel-talk fashion, but obviously in a tone
favourable to Romanian claims in this area.

The influence of geographers at Versailles is hard to assess, but
apparently considerable: Jean Gottman[44] has commented that
many features of the new map of Europe, drawn at the Versailles
peace conference of 1919, particularly the boundaries of Romania
and Poland, owed much to the friendly co-operation between
Bowman and de Martonne, but the difficulties of defining both are
apparent in Bowman's *New World*. D. W. Johnson,[45] in an article
on 'A Geographer at the Front and at the Peace Conference',
mentioned the work of the commissions appointed to consider the
territorial claims of various existing and potential states. After

various deputations had been received, the members of the commission 'and their associated geographical, economic, historical, military and other experts would debate the issue at length, and decide what was just in each claim and what was unjust, and where the new boundary lines should be drawn, striving to fix the frontiers as nearly as possible along the lines of racial division but taking into due account the geographic, the economic, and to some extent the strategic factors, in order to get the wisest and most permanent settlement of the various complicated territorial problems'. Johnson adds that the advice of the territorial experts was frequently sought and extensively used. Certainly there had been no lack of preparation, for as G. M. Wrigley shows in her appreciation of Bowman, from 1916 all the resources of the American Geographical Society had been placed at the government's disposal and the 'Inquiry' on the geographical bases of a political resettlement employed some 150 experts. So much learning came over the Atlantic that the boat conveying it has been described as 'groaning and creaking with erudition'.[46]

So far the full story of the making of the Versailles Treaty has not been told, in spite of the publication of numerous memoirs and more serious works. But ethnic maps became sources of argument as never before or since that time, especially as several conflicting versions might be available for the same area. Bowman[47] has described how the boundaries of the Free City of Danzig were drawn: when Lloyd George and President Wilson had agreed to make the Free City, he and the British delegate, Mr. Headlam-Morley, 'decided to avoid discussion of the relative merits of the ethnic maps of the different delegations by submitting a small map prepared by Mr Lloyd George's advisers.' Then 'Mr Paton of the British delegation and I set to work upon a large-scale map prepared by the American Inquiry, which was used throughout the Polish negotiations as the authoritative map on ethnic matters. Between four and six o'clock we traced the boundaries of Danzig as they stand in the treaty today. Transferring these boundaries to the British small-scale map for the benefit of Mr Lloyd George they were presented to the Council of Four and there passed without delay'. Bowman served on many commissions as a representative of the United States as well as on the Polish-Ukranian Armistice Commission; after the Versailles treaty was signed, he worked for the Central Territorial Committee in Paris on the treaty between the Free City of Danzig and Poland, the plebiscite in Cieszyn (Teschen), in negotiations with Bulgaria and on the problems of Italy and Yugoslavia in the Adriatic. Partly from the contracts made in Paris, two famous societies were founded—the Royal Institute of International Affairs in Britain and the Council on Foreign Relations in the United States—each of which has a distinguished record of publication.

One main preoccupation of post-1918 Europe was to prevent the alliance or union of Germany and Russia by establishing a zone of strong buffer-states between them. As the Austro–Hungarian monarchy had become weaker in the years before 1914, so the fear of German expansion grew and various geographers spoke of Mitteleuropa, not necessarily with political or imperialist overtones, though these were required in time. In a review of these ideas, H. C. Meyer[48] says that before 1871, the term Deutschland was used as an ethnic and geographical description for the German states and Austria: before 1914 the term was generally identified as the historic German lands, though there were wide conceptions and implications: in 1883, for example, Emil Deckert[49] described Germany as in a position to dominate the transport, trade and cultural life of its neighbours, Austria–Hungary, Switzerland and the Low Countries. But the most famous geographical treatment of Mitteleuropa came with Partsch, who included an area from Ostend to Geneva and from Memel to Burgas, along the Alps to Trieste, along the Balkan mountains north to the Danube delta and north-west along the political boundaries of Roumania, Austria–Hungary and the German Reich. Physically, said Partsch, Mitteleuropa included three elements, the Alps, the middle mountains and the northern lowlands but wherever one of the elements vanished, Mitteleuropa came to an end. Whatever use has been made of Partsch's conception of Mitteleuropa, it was intended primarily as a physical basis for regionalization; another famous German geographer, A. Hettner in 1907, excluded from Mitteleuropa the middle Danube basin, the lower Danube valley and all of the Balkans, to make south-eastern Europe.[51]

Geography appeared to give some kind of sanction to the continental imperialism of Germany: in 1919, for example, Commander Roncagli[52] said that 'German geographers have long ago tried to make physical geography one of the moral weapons with which Germany prepared to carry out her plans of dominating the world'. He then deals with A. Penck, who in 1916 argued that the natural boundary between central and southern Europe lay at the foot of the southern /sic/ slopes of the Alps. Partsch spoke of the area where German was understood as 'everywhere from Galatz, Sofia, Sarajevo, Trieste, Genoa and Antwerp far into the interior of Russia. Only the most backward regions of Serbia and Montenegro must be excepted. All the rest of Central Europe, consciously or unconsciously, belongs to the sphere of German civilization'. In time the conception of a German-controlled central Europe went further and Penck spoke of Zwischeneuropa, between Vordereuropa (Norway, the British Isles, France and Iberia) and 30° east, beyond which was Hintereuropa. In the 1939–45 war many comparable ideas existed, and a probable development after a German victory would have been some form of Sudosteuropa under German control, of which the economic arrangements with Balkan countries before 1940 were the first symptoms. Since the war, as E. Fischer[53] has pointed out, there is politically no Mitteleuropa or Zwischeneuropa, for the Iron Curtain has divided Europe into only two parts, west and east; and the buffer between Germany and Russia has become a line of states looking east with, however, two notable exceptions, Finland and Yugoslavia.

The concentration on Europe in this chapter does not imply that there are no aspects of political geography worthy of study elsewhere. Indeed the author was first led to some interest in political problems by work on the population of China and Japan, and the emigration from these countries, notably to the East Indies and in the case of Japan to various mandated territories of the Pacific where the number of migrants seemed remarkably large. And a fascinating study lay in the position of Manchuria, the last of the world's great grasslands to be thoroughly settled, with Russians in the north, Chinese everywhere and Japanese scattered around in relatively small but highly influential numbers. Throughout the world there are political problems which demand geographical attention: even within the British Isles, to the author's regret, no one has yet been found to make a thorough study of the border between Northern Ireland and the Republic of Ireland; and one would welcome a detached study of the present apparent redistribution of population in South Africa. On the other hand, there has been excellent work done since the 1939–45 on European population movements, much of it in geographical journals. And along with this, one must mention the work of economists, contemporary historians and others for whom, perhaps, the international scene changes all too quickly.

Geopolitics

An intelligent German research worker in 1938 told the author that Czechoslovakia could not possibly survive as it was such a peculiar shape. Such comments perhaps show the intellectual level of geopolitical thinking, though fundamentally its subject of study was the theory of the state and the relation of people and their economic activities to a country, all of which are obviously worthy of consideration. The trouble was that geopolitics rapidly became a *Weltanschauung*, obliged to be practical, to point the way to the future, to plan and devise. Originating in Sweden with Kjellen (1864–1920), geopolitics attracted little attention until the 1914–18 war, when Kjellen's main work[54] *Staten som Lifsform* appeared in 1916. The German traveller and soldier, K. Haushofer (1869–1946) considered political geography to be concerned with the distribution of political power, conditioned by and dependent on surface features, climate and use, that is, an academic study; but geopolitics was dynamic as it furnished the implements for political action and was a guide to political life—as it were, the geographical conscience of a state. Geopoliticians accepted from Ratzel the idea that the state was an organism[55] and some went so far as to say that states go through stages like any other organism: the soil and men are inseparably bound together, and states have a compact kernel or core area (cf. p. 215) and a looser structure in the tributary area, finally dissolving in a series of spearheads in alien territory.

According to circumstances a state might develop within its historic boundaries or embark on a career of conquest; ideally the state should become a natural unit, occupying a geographical unit such as a region, and in strong states expansion might be necessary for self-preservation, as in the case of Japan. It was easy to show the need for territorial expansion, for *lebensraum*, on such an argument, which incidentally rests on the view that in any country the population should be increasing. Before the 1939–45 war, many Germans regarded France as decadent because its population was virtually stationary, while that of Germany was increasing by half a million a year because the birth-rate had increased since 1933, largely due to family endowment schemes and government propaganda.[56] The consideration of shape seems odd, and no doubt the German research worker regarded the inclusion of Bohemia–Moravia and Slovakia within the German

Reich in 1939 as scientifically inspired, since it removed an awkward element from the map of Europe and gave the Reich something nearer to the circular or square shape regarded as good. Incidentally, some geopoliticians regarded a long elongated shape such as those of Norway, Chile and Italy as undesirable, yet how can these countries help it? Especially as Norway's separation from Sweden is never in question by either? Yet in such countries there are obvious problems of communications.

It is not the author's intention to pour scorn on geopolitics. The trouble is that its twentieth century development has brought it into bad repute: much of its basic material is indistinguishable from that of political geography, together with political science and history, and the geography has been used to give a pseudoscientific flavour to various theories of political planning. Mackinder's work on the heartland was taken up by various geopoliticians to illustrate the permanent threat to world peace of Russia and the necessity of an anti-Communist crusade. A geographer may reach certain conclusions on the distribution of world power through his study, but once he begins to advise action, he becomes either a politician or a planner. And this is not in itself undesirable.

CHAPTER TEN

THE ADVANCE OF CARTOGRAPHY

Maps from firms and individuals; atlases of the nineteenth
century and later; national atlases.

MUCH of the story told in the previous chapters has been
concerned with the mapping of newly acquired data and,
as shown in pp. 233–9, the modern advance of regional
geography was made possible by the publication of various atlases
that dealt with distributions both on a world view and for countries
or smaller areas. Explorers provided initial surveys; exact
measurements by national surveys followed; and eventually all
kinds of physical and social distributions could be mapped. A
century ago, Guyot in America was measuring the heights of
mountains; rather later skilled and imaginative geologists were
making surveys that laid some of the foundations of modern
geomorphology; more recently geographers have produced fine
atlases of the agriculture and historical geography of the United
States. In modern field-work the surveyor may go forth with a
map on the scale of six inches to one mile to map rural land-use,
and with a twenty-five inches to one mile map in towns. One major
advance has clearly been the increased intensity of the work done.

The phrase 'maps are the tools of the geographer' is a cliché,
and a foolish one. The geographer is by no means unique in his
use of maps for illustrative purposes, as they are used by all kinds
of people—archaeologists, historians, political scientists, botanists,
geologists, zoologists, planners, government administrators and
many more. Whether such workers necessarily use map evidence
and presentation to the best advantage one may doubt: many
historians, in spite of the immense advance of their scholarship in
the past hundred years, seem strangely reluctant to use the
evidence that may be acquired from old maps or even to arrange
for their work to be illustrated by maps at all. The work of E. A.
Freeman[1] (see p. 207) was based partly on map compilation
and it may be added that geography was at one time studied in
many honours schools of history; why then should historians be so

reluctant to study maps from the nineteenth century with the care they would devote to documentary evidence? Portraits of Victorian scholars not uncommonly included a globe as an appropriate background, and polite nineteenth-century education, even for girls, was incomplete without the 'use of the globes'.[2] Every map is, in greater or less degree, a compromise picture of the spherical world, and the study of map projections has been a source of fascination to many people; many ingenious examples are given in atlases. Always the right choice has to be made. For purposes of distribution mapping, equal area projections are obviously essential, whereas if accurate direction is needed the Mercator projection is the best choice, though it distorts areas increasingly towards the Poles so that Canada is made to appear many times as large as the United States.

A survey of the whole field of cartography shows that the main debt, historically, is owed to the private producers and more recently, effectively from the nineteenth century, to the various national surveys, many of which have been connected in some way with national defence. While much has been written on the history of maps,[3] there is still a great deal of material available which would repay attention; and biographical inquiry into map-makers, though by no means neglected, would be rewarding. The successful map-maker was a product of his times, meeting a need and arousing interest, and generally, indeed, obliged to do just that if his maps were to sell. Enough has been said in previous chapters to establish the point that the last hundred years gave abundant opportunities to the makers of maps: not only was the world being opened to men's eyes, but they were avid to see and know the expanding world for themselves. The readers of geographical journals a century ago eagerly studied not only the accounts of strange lands that filled their pages, but also the somewhat tentative maps that accompanied them.

Maps from Firms and Individuals

Maps have been provided by individual cartographers for centuries, though the history of their efforts lies beyond the immediate scope of this book. But as A. H. Robinson[4] has noted, 'during the period of recorded history, from Babylonian times to the present, cartography and geography have been intimately associated . . . many of the earlier geographers might more appropriately be called cartographers.' It was appropriate that the first international congress of geographers[5] was a festival at Antwerp in August 1871 to celebrate the work of the great Flemish geographers, Mercator and Ortelius, to whose memory statues had been erected by public subscription at Antwerp and Rupelmond. This festival was termed the *Congrès des Sciences géographiques, cosmographiques et commerciales*, but it was agreed that the commercial element must be distinctly geographical. The whole development of geography has been associated with cartography, and most maps are an accumulation of past experience, depending on the surveys and levellings of previous workers, except in the case of an initial or an entirely revised survey. This makes foolish such claims as those given for a recent new atlas in Britain that 'every map begins with a blank sheet of paper'; of course it does, but what goes on the paper owes much to the workers of an earlier time, with refinements and renovations. Everything is drawn that it may be re-drawn later on. In 1817, A. Petermann[6] noted, 'Humboldt based his isothermal lines upon observations made in sixty places; in 1832 Kaemtz, of Dorpat, had increased the number to 145; in 1839, Berghaus, of Potsdam, published a list of 307; and in 1844, Humboldt, in his work on *Central Asia*, gave as many as 422.' Petermann's figures show how inadequate the statistical basis of such isothermal maps was—it still is in many parts of the world.

Maps of cities and towns by private cartographers have been numerous. There are, for example, some fine maps of London[7] by J. Rocque, a Frenchman who settled in London, whose work belongs to the period 1734–62; and in the early nineteenth century E. Mogg provided more fine maps, including in 1813 a plan of the new Regent Street described as 'reduced from the Large Plan in the House of Commons'. Many of the government reports of the time were excellently illustrated by maps: for example, the three-volume report of the 'Commissioners . . . upon the boundaries and wards of certain boroughs and corporate towns in England and Wales', published in 1837, contains a fine series of town maps on the scale of four miles to one inch. Each map shows the various wards of the town in hand-applied colour, and in the preface it was noted that the Circuit Commissioners had visited the boroughs and done their work with zeal, skill and industry. The responsibility for the maps[8] lay with Lieutenant R. K. Dawson, of the Royal Engineers, whose co-operation was valued 'not only in the preparation of the Plans, but in the revisal and final recommendation of the Boundaries and Wards'. Thanks were given also to Lieut.-Col. Colby (p. 45) 'for the assistance . . . derived from documents of the Ordnance Survey, from which the accompanying plans have been for the most part, prepared'.

E. W. Gilbert[9] has drawn attention to various maps of the nineteenth century which were produced to show the distribution of cholera cases: these included a 'cholera plan' of Leeds in 1833, with the afflicted areas coloured red, and others of later date for Exeter, Oxford and London. A map of 1855, by Dr John Snow, showed that the 'cholera field' was in an area between Regent Street and Dean Street, and he traced the cause to infected water at a pump (p. 197). His map, on the scale of thirty inches to one mile, was included in the second and enlarged edition of a paper 'On the mode of communication of cholera', published in 1855: it plots the distribution of deaths by black rectangles and also marks the pumps. On a national scale, A. Petermann produced in 1852 his 'Cholera map of the British Isles showing the districts attacked in 1831, 1832, and 1833'. Gilbert has noted that the *Physikalischer Atlas* of Berghaus, published from 1837–48, included a map of the geographical distribution of diseases: the second edition of A. K. Johnson's *Physical Atlas of National Phenomena*, published in 1856, included one plate showing the world distribution of diseases and another showing the march of cholera from east to west, with dates of its occurrence.

The maps discussed above had an obvious practical significance; and many more of the same period are not less informative and helpful. In a recent paper, A. H. Robinson[10] has said that 'the period 1835–55 might well be termed a "golden age" of the development of geographic cartography'. He illustrates his statement by a discussion with illustrations of the maps prepared in 1837 by H. D. Harness for the Second Report of the Irish Railway Commissioners in a special atlas. This had a geological map, maps of Ireland on the scale of four inches to one mile, and—perhaps more remarkable still—maps showing population density and the flow of traffic. The population map makes an excellent attempt to exclude those extensive areas of Ireland which were unpopulated, and grades densities elsewhere by aquatint shading in three tones: unfortunately these are not entirely clear, but the densities per square mile of the rural population were calculated for each barony (into which the counties were divided) and placed on the map. The two 'traffic flow' maps employ a line shading varying in thickness according to the amount of movement: one deals with public conveyances and the other with the general conveyance of merchandise. In the Census of Ireland[11] of 1841, there was a population map with a line shading in five tones, but no effort was made to exclude the unpopulated areas; a second map shows the percentage of houses consisting of only one room, a third the prevalence of illiteracy, and a fourth the relative value of livestock. This map was at once ingenious and naive, for there is no such thing as a standard horse or cow in value. Taken as a group these maps illustrate the difference between east and west which is crucial in the geography of Ireland, but it is unfortunate that Harness's excellent precedent of excluding unoccupied areas from the population map was not followed. Even more remarkable, perhaps, was the inclusion of a map of Dublin with some interesting data on the quality of its main streets, including those of a residential character. Colours inserted by hand are used for a six-fold classification, first- and second-class 'private streets', first, second and third classes of 'shop streets', and 'mixed streets' of the third class, which means very poor streets. The subdivision cannot have been easy to make, and one hint of the difficulty is given by the use of different colours for the two sides of the same street. The map is a forerunner of the qualitative mapping in the famous surveys edited by C. Booth,[12] published in London in 1892 under the title *Life and Labour of the People of London*, in which the streets of London were given coloured symbols ranging from black, for the 'lowest class . . . vicious, semi-criminals' to yellow, for the 'upper-middle and upper-class families . . . wealthy . . . with three or more servants' and 'houses rated at £100 or more'.

A survey of Manchester,[13] now of considerable historical significance, was published as Adshead's Twenty-four illustrated maps of the Townships of Manchester divided into municipal wards corrected to the 1st May 1851. It is on the scale of eighty inches to one mile, and is described as 'an original survey by Richard Thornton, carefully corrected April 24, 1850'. In effect, this map is a thorough survey of land-use as it shows public buildings; warehouses and other places of business; hotels, inns and public houses; private houses; mills, works and yards—these last are uncoloured but the premises are named. One interesting feature is that the warehouses cover much the same areas as at present, and the detailed survey of housing shows exactly where the courts and back-to-back houses were located. The width of street is shown, and a note is given of those paved. The general effect of this black-and-white map is attractive.

If, as suggested above, the period 1835–55 was a golden age of cartography (but not necessarily the only one), there were many circumstances encouraging to the makers of maps. The growth of government mapping in Great Britain, Ireland, and many other countries gave a basis for further work, and once railways were planned, exact surveys and recording of observations on maps proved necessary. The earlier surveys for canal construction and road building, or at least road improvement, had been illustrated by finely-drawn maps; as an art, and a very delicate, finely-culti-vated art, cartography is of immense antiquity. But at the time, rather more than a century ago, when new transport was opening up the world, there was an inevitable stimulus to cartography. There was also a growing interest and concern in the state of the towns and cities in which an increasing proportion of the popula-tion were living, and it is not remarkable that some maps came from medical sources: in Britain, for example, government reports such as that on the Health of Towns[14] showed an acute apprecia-tion of the problems that had arisen or would be likely to arise, and when various boards of health or other bodies were set up to deal with what were then called 'sanatory' conditions the need for new and improved maps became greater. Anyone doubting this state-ment would be well advised to consult some of the numerous government reports that were issued at the time, or some of the initial surveys of the railways available in libraries. Privately-printed gazetteers such as Samuel Lewis's *Topographical Diction-ary of England*[15], published in 1831, include maps of towns; in this volume it is commented that 'the plans of the counties, cities and boroughs have been drawn and engraved with great care, and were afterwards submitted for examination to the different returning officers, to whom we offer our grateful acknowledgements for the promptitude with which they uniformly replied to our requests'. There are in the Lewis volume 116 plates, most of which include two or more towns with the borough boundaries of the time: the maps were 'engraved by R. Creighton' and 'engraved by J. and C. Walker'. These maps bear an interesting relation to those in the report of the Municipal Corporations commissioners noted on pp. 228–9.

Petermann's work included two fine population maps based on the 1841 and 1851 census. The first of these was exhibited at the

Statistical section of the British Association for the Advancement of Science at Swansea in 1848 and published in 1849; it was not, as A. H. Robinson[16] said, the 'second shaded map in existence', as it had at least two predecessors, the Harness map of 1837 and the Irish Census map of 1841. Like the Harness map, Petermann's 1849 map[17] excluded the unpopulated areas, and showed population densities by a graduated shading, with symbols for towns of various sizes: these were a small circle for towns of 3,000 to 10,000, a small green circle for those of 10,000 to 20,000, an orange circle for 20,000 to 50,000, a black square for 50,000 to 100,000 and a hexagonal figure filled with red for the largest towns, over 100,000. Petermann was also the author of the population map in the Census of Great Britain,[18] 1851, which has the darkest shading for densities of 600 persons and upwards to a square mile, figures showing the average density for each county, and black spots, of graduated sizes, representing all towns with more than 2,000 inhabitants. Petermann[19] produced other population maps, including one for the National Society (for the education of the poor) in 1851 which showed the size of towns with over 5,000 people by black spots of graduated size and, by number in thousands of the populations for each county and the larger cities. Later, Petermann[20] employed a similar technique in a map of Spain published in 1856. The custom of including statistical maps in censuses survived for a time: for example in Ireland the 1861 Census had three maps of population density, for 1841, 1851, and 1861, placed side by side, and in 1881 the comment[21] was made that 'the uses of maps and diagrams in illustration of statistics are now so universally known that it is unnecessary to refer to them here'. At this Census, there was a set of maps in colour showing population density, the percentage of illiterates and the proportion in accommodation of the fourth class, or poorest type. In 1891, another map, showing the valuation per head was added, but with the refinement that it was calculated for each poor law union, instead of merely for each county.

Several firms and individuals produced splendid maps at this time. John Arrowsmith prepared a number of maps for the journal of the Royal Geographical Society.[22] The work of the Johnston family—firm at Edinburgh is mentioned on p. 45 and its relations with the Berghaus cartographical school noted on p. 60. The famous Perthes firm was established at Gotha in 1785, and the success of its work was said—in 1885—to be due to its use of 'all the best geographical talent in Germany'.[23] But the real advance depended on the work of explorers: as C. R. Markham[24] commented in 1880, 'in 1830 only unconnected strips of the coast of Arctic America (were known); knowledge of the east sides of Greenland and Spitzbergen, the coasts of Novaya Zemlya, and the surrounding seas was vague and inaccurate—an enormous area was totally unknown. Now, the whole coast of Arctic America was delineated, the remarkable archipelago to the north explored, and seven north-west passages worked out: Nordenskjöld has achieved the north-east passage.'

A further comment was that a comparison of the atlases[25] of 1830 and 1880 showed not only the results of discovery but also 'great improvements in the methods of investigation, in systematic arrangements of facts, in cartography, and in the construction and use of instruments'. One example, compiled from a variety of sources, is F. F. von Richthofen's *Atlas for China*,[26] published in Berlin in 1885: it included twenty-seven physical and twenty-seven geological maps. The coast was taken from Admiralty charts, and the surveys of Jesuit missionaries published at Wuchang in 1863 were also used; von Richthofen notes that the topographical features were doubtful in places, and for example, some symbols 'only indicate the general direction of the ranges'. Heights were measured by the aneroid barometer, and the geological maps were compiled from the author's own observations with any other available sources. The long papers of an exploratory nature may be boring now, but at the time they were a considerable addition to knowledge: A. Hosie,[27] for example, added much to the knowledge of Yunnan and Szechwan and provided valuable data for the maps of China, both in his papers in the journal of the Royal Geographical Society and in official government reports. He was H.M. agent at Chungking, which was in 1886 already regarded as 'the greatest commercial centre in Western China'.

Atlases of the Nineteenth Century and Later

These are of great variety, and naturally increasing complexity. An early example is Ritter's maps of the physical geography of Europe, published in 1806 (see p. 32), followed six years later by an atlas from the other classical figure of the time, A. von Humboldt. His atlas, published in Paris, as 'Atlas géographique et physique du royaume de la Nouvelle Espagne, fondé sur des observations astronomiques, des mesures trigonomêtriques et des nivellemens *[sic]* barométriques', included hachured maps of central America, and a map of the United States which, in what is now Louisiana, has the comment 'Plaine immense où passent les Bisons'; there are other vast empty spaces. The atlas[28] includes a plan of Vera Cruz, and two section-profiles, from Mexico to Acapulco, and from Mexico to Guanaxata; von Humboldt was greatly interested in the possible construction of a canal from the Pacific to the Atlantic. From this time, a rough distinction could be made between atlases that were primarily topographical and those largely concerned with distributions; the first edition of Stieler's *Hand-Atlas*, published by Perthes of Gotha, which appeared in 1834, began the long series of topographical atlases published under this title.[29] Like many more 'Hand' atlases, it is of vast size and weight.[30] Some atlases, however, are partly topographical and partly concerned with distributions, and a number have had some special purpose, such as one published in 1842 as 'The Colonial Church Atlas arranged in dioceses with geographical and statistical tables', which showed the Anglican diocesan organization over the world.[31]

Berghaus's atlas[32] was first published from 1837–48, but a revised edition, with more than ninety maps, appeared in two volumes (1849 and 1852). The first volume had four sections— meteorology and climatology, hydrography, geology and terrestrial magnetism. Many of the maps in this first volume were naturally incomplete and even speculative; the maps in the second volume, on plant geography, zoological distributions, anthropography and ethnography are of considerable interest. The relation of isothermal lines was shown as far as the data of the time allowed, and the heights reached by vegetation in such areas as the Himalayas, the Pyrenees, the Alps, the Andes and—interesting to note—the island of Teneriffe, were shown. The zoological section is illustrated with vivid, even alarming, representations of the animals concerned, and there are also illustrations of Turks, Chinese and others to accompany the distribution maps. Linguistic groups are shown, and an interesting map shows where German was spoken, and suggested that much of Sweden had the Lapp language; in Ireland, *Ersen* was recorded as spoken everywhere except in a narrow strip near Dublin, apparently the English Pale, and the four northern counties of Antrim, Down, Derry and Donegal. The atlas includes political maps and one of the native American tribes. There is a lengthy text, and later editions appeared in 1886 and 1892.

In 1843, some of Berghaus's maps were published in the 'National Atlas of historical, commercial and political geography' by A. K. Johnston of Edinburgh.[33] In the preface, Berghaus quotes a statement of A. von Humboldt to the effect that 'graphic' methods of presenting the data of 'natural philosophy' give a convincing impression to the eye. Berghaus acknowledges his debt to 'German and British naturalists' [sic], and offers to the friends of geography in Britain four sheets of his 'Physical Geography . . . larger and more complete than in the German edition' (of his atlas). The first plate shows Humboldt's isotherms, with the limits of Arctic and Antarctic pack-ice, and graphs of the 'hourly medium temperature of the year in the temperate zone', with two examples, for Padua and Leith, 1824–7. Next follows a map of the main food crops of the world, including cereal-corn, wheat, rye, maize, barley, sugar-cane, coffee, and—rather oddly—vanilla and spices. In another part of the atlas, the world distribution of barley, oats, rye, wheat, maize and rice is shown, with special symbols to represent combinations of crops such as barley-oats, rye, barley and wheat, rice and maize. Plate 1 includes the 'isotheres' or lines of equal summer temperature, with the thermal equator, 'the line of maximum temperature of the air'; also shown is the general and the occasional limit of ice in the Arctic, to the south and east of Iceland, and the summer ice around Spitzbergen. The third map is titled 'geographical distribution of the currents of air, perennial, periodical and variable winds, and regions of prevalent hurricanes': it shows the seasonal movement of the trade winds, and implies that they are permanent between the latitudes of 25°–9° N. and 4° N. to 23° S. Both to the south and the north there is the germ of the Polar Front theory, evolved some seventy years later; what became known as the 'westerly wind belts' is referred to (in the southern hemisphere) as the 'region of north-west currents of air or of the downward returning South-Eastern Trade-Winds in triumphal conflict with the Southern Polar Currents'. The monsoon winds in the Indian Ocean are also indicated, and hurricanes are shown in relation to North America and by one solitary arrow reaching Madagascar. The map is based largely on marine observations of the Prussian merchant navy and the English hydrographers. A fourth plate is primarily physical, and includes a map showing the lines of mountain chains in Europe and Asia with a geological map of Java; this plate also includes an essay on the heights of the continents by Humboldt, with illustrative diagrams.

Study of the previous paragraph will perhaps reveal to many students of geography that many familiar maps are, in fact, derived from sources of some antiquity: as noted on p. 34, Vidal de la Blache rightly hailed Berghaus's work as a definite advance in geography. The 1843 Johnston atlas also includes an 'ethnographic' map of Europe by Dr Gustav Kombst, which has a revealing sub-title, 'different nations of Europe traced according to race, language, religion and form of government—in which the physiological, moral, and intellectual character of the Celtic, Teutonic and Sclavonic [sic] and other races of men, is graphically described.' Dr Kombst was certainly sure of himself. He includes the whole of the Iberian peninsula as Moorish south of a line drawn a short distance south of Lisbon to Valencia, and gives full recognition to the Celtic fringe of Europe by distinguishing the areas of Gaelic, Welsh, Cornish, Erse and Breton speech. In Ireland, the map strongly resembles that in the German Berghaus atlas (p. 234). The accompanying text has frank comments on the qualities of the peoples of Europe which make interesting reading: the Celtic variety, for example, is 'irascible, not forgetful of injuries (with) little disposition for hard work, bad seamen, and not fit for colonizing'.

The Johnston firm, in 1850, published a *Physical Atlas of Natural Phenomena* which included some maps derived from Berghaus but was in fact an independent production.[34] Like most such atlases of its time, it included a lengthy descriptive text, and the maps fall into four main categories, geology, hydrography, meteorology, and natural history. The physical section includes trend lines for the mountains and a large number of sections, and it is noted that the mean heights of the continents had been 'elucidated in an interesting memoir by von Humboldt'. There are five plates on animal life and one on ethnography, but perhaps the most interesting section is on climate, which included Humboldt's isotherms 'with lines of equal barometric pressure at the level of the sea'. Annual average temperatures are shown also, with figures for winter and summer, and there are maps showing isotheres, or lines of equal summer temperature, and isochimenes, or lines of equal winter temperatures. The map of winds bears a clear resemblance to that in the 1843 atlas noted above, but in the caption about the relation of polar and tropical (trade-wind) air, the word 'triumph' has been removed and only the 'conflict' of these two great airstreams remains. There is a hyetographic or rain map of Europe, showing the number of rainy days (more than 0·01 inch), the isohyetoses, or lines with equal amounts of rain, now commonly called the isohyets, and isotherombroses, the percentage distribution of summer rain. In the section on plants, the altitudinal trends of particular species receive attention. This atlas was soon sold out, and a second edition, published in 1856, of 2,500 copies was also well received and even praised by Ritter.

Meanwhile A. Petermann[35] had produced an *Atlas of Physical Geography*, published in 1850; he was described as 'formerly of the geographical establishment at Potsdam, and many years assistant in the Great Physical Atlas of Berghaus'. His atlas, 'illustrated by one hundred and thirty vignettes on wood', included a 'descriptive letterpress by Rev. Thomas Milner, M.A., author of *The Gallery of Nature*, etc. etc.'. The atlas followed much the same plan as Berghaus, with plates showing the distribution of active volcanoes, world hydrography, currents and temperatures, a world meteorological map distinguishing zones which were 'torrid' (over 70° F.), 'temperate' (30° to 70° F.) and 'frigid' (under 30°), maps of winds and rain distribution—and also, characteristic of the time, plants, mammals, birds and reptiles and ethnography. There was a special physical map of Palestine, and maps of the British Isles, under the headings orography, climatology, botanical and zoological.

Almost all the atlases so far considered were concerned mainly with the major distributions of climate, flora, fauna and man over the surface of the earth. Their primary aim was to give a world view, but their compilers were not unmindful of the local deviations, for example in mountainous areas. That such atlases laid the foundations of a world regional geography need hardly be stressed; no doubt many readers of the last few pages may recall its similarity to what has been taught in the classrooms of schools, colleges and universities in general introductory courses. It will be noted, too, that the tendency to proliferate technical terms is by no means confined to the twentieth century. Along with such atlas there was also a continuing publication of others that were predominantly topographical and political, such as Stieler (p. 234), which reached its tenth edition in 1930, though some of the previous editions had numerous revisions—for example the ninth edition, published in 1905, had five separate revisions. Another example is Johnston's *Royal Atlas* (p. 60), designed to show places and political divisions, which went through twelve editions between 1859 and 1908. Modern atlases of the same type include the *Times Atlas*, published by the Bartholomew firm of Edinburgh in 1920, and including the post-1914–18 war settlement, and its completely revised successor,[36] published in five volumes from 1955–9. Comparable atlases include that published by the Touring Club Italiano from 1929 and the Swedish *Nordisk Världatlas*,[37] 1926. These atlases contain a varying number of world maps, but their main aim is to show individual countries with physical features and places. For Scotland, a notable example is the atlas sponsored by the Royal Scottish Geographical Society and published in 1895 by J. G. Bartholomew, which has the whole country on the scale of two miles to an inch in layer colours, a technique pioneered by this famous map-making family.[38] There are also smaller-scale maps of the physical features, geology, climate and natural history, largely compiled by A. Geikie and A. Buchan. The last-named, a well-known meteorologist of his day, also edited the *Atlas of Meteorology* which appeared in 1899 and included over 400 maps of climate and weather. This was Vol. 3 of a projected *Physical Atlas* in five volumes[39] by J. G. Bartholomew, but only this volume and one other—Vol. 5, Zoogeography, 1911, appeared. Among the staff at the Bartholomew headquarters at one time was A. J. Herbertson (pp. 83–4), and the influence of this phase on his later work on natural regions was clear.

One atlas that has had a considerable impact on the teaching of geography was published in Paris in 1894 as the *Atlas Général Vidal-Lablache*[40] with later editions in 1909, 1918, 1922, 1938 and 1951. Although the 1951 edition belongs to an age very different from that of 1894 and has been brought up to date, the appearance, and even the size, of the two atlases is much the same, though the maps are more numerous: the 1951 edition has 385 maps on 130 plates and the original work 248 maps on 131 plates. The summary notes dealing with each map have been revised by various people, including E. de Martonne, the geographer who became the leader of the French school after the death of Vidal de la Blache in 1918. The original preface of 1894 pays homage to the pioneer mapping of Ritter in the six maps dated 1804–06, and explains that for each country the political map is accompanied by a physical map, as each illumines the other; further clarification comes from geological, climatic and statistical maps. The fundamental idea is one of relationship between the life of a country and its physical environment. Study of this fascinating atlas will reveal that it gives a panorama of world history, and incorporates many interesting maps showing past political boundaries: just as several of the atlases already mentioned gave a world view of climate, vegetation, physical features and the like, so this fine atlas, of a cartographically most attractive type, did something to further the idea of the relation between world history and geography.

The *Atlas of the Historical Geography of the United States*[41] appeared in 1932. It includes an explanatory introduction and index, with a comprehensive series of maps; to a great extent it makes possible the oft-quoted ideal of historical geography—to reveal the regional geography of the past. After a useful if conventional group of maps showing physical features, soils and vegetation, the area of virgin forest in 1620, 1850 and 1926 is shown, with the national forest of 1930. Another set of maps shows climatic data, including the length of the growing season, and the distribution of oil and minerals. There are reproductions of historical maps from 1492 to 1867—a fascinating section—which largely tell their own story: among them are maps by Mercator 1569, Ortelius 1589 and among more modern cartographers von Humboldt 1811 and Arrowsmith 1814. Several plates are given to the Indian tribes, with their distribution in 1650, the battlefields of various periods, and the reservations, past and present. Explorers' routes are shown and full attention is given to various international and state boundary disputes. The territorial expansion of the States from 1783–1853, and the extent of the public land at twenty-year intervals from 1790–1910 is mapped. A complementary series of maps shows the areas settled at various dates with the towns of the time. Social distributions receive considerable attention, and include the proportion of slaves in each county from 1790 at ten-year intervals to 1860: the number of free Negroes is shown in each county—not, it should be noted, as a percentage of the population. There are maps showing the distribution of the coloured population, and of the main sources of immigrants to the States from 1831–40 onwards, with the percentage of foreign-born at various periods, and the number per county of Germans in 1880, Irish in 1900 and Swedes with Norwegians in 1930. The density of population is shown for 1790 and every ten years to 1930. Other social maps include the distribution of colleges and universities, of churches of different denominations and even of the voting at elections. An historical series of maps shows the iron works of 1620–75, and the iron and steel plants of 1725–75, 1858, 1878 and 1908; cotton-spinning mills are given for 1810, 1840, 1880, 1926 and there is a cartographic treatment of communications, of imports and exports from 1851–60 onwards, and of agricultural regions and particular crops. Finally a number of historic town plans are given. One is tempted to lapse into superlatives in speaking of this atlas, and particularly to wish that from Britain one had a comparable range of information within the covers of one volume.

National Atlases

These are numerous but varied in quality:[42] several have been published by geographical societies, and possibly a claim could be made that the Bartholomew Atlas of 1895 for Scotland (p. 238) was the first truly national atlas. In many countries no officially recognized national atlas exists: in Great Britain, for example, the series of maps on the 1:625,000 scale published by the Ministry of Town and Country Planning (later by the Ministry of Housing and Local Government) serve much the same purpose as a national atlas. They include some interesting maps of physical features, geology, administrative areas, land-use, farming, population density and movements, economic minerals, power distribution and industrial location. Many of the maps are of direct significance for planning purposes, and as there is constant replenishment of the series, some are already of historic interest. Some national atlases, such as that of France, have appeared as loose-leaf sheets to be incorporated into a cover; unfortunately in Britain the 1:625,000 scale is too large for binding to be practical. At one time the Ordnance Survey announced the intention of issuing these maps on half the scale, so that an atlas could be made; many of the maps could, in fact, be reduced to this scale without the loss of clarity. It is not for want of trying: both before and after the 1939–45 war, strenuous efforts were made to achieve a national atlas, largely by a British Association for the Advancement of Science Committee and the Royal Geographical Society; one name must be mentioned for particularly devoted work, Eva G. R. Taylor. In 1961, however, the Oxford University Press announced the publication of an atlas corresponding in scope to several of the national atlases.

Equally, it could be urged that the fine American work mentioned above has something of the character of a national atlas. At least it shares the avowed aim given in the Finnish atlas, 'to assist the people of Finland to know themselves and their country'. One might assume that a national atlas would be found not only in libraries but in the homes of cultivated and thoughtful citizens. The existing atlases vary in quality and a full analysis of them is beyond the range of this book; in any case actual acquaintance is far more rewarding. But a check through the catalogue in the Royal Geographical Society revealed that national atlases are numerous: the following, not necessarily a complete list, includes the dates of later editions and new atlases: Finland 1899, 1911, 1925; Sweden 1900, 1953; Canada 1906, 1915, 1959; Gold Coast 1928, 1935, 1945; Egypt 1928; French Indo-China 1928; Czechoslovakia 1935; France 1933 onwards; U.S.S.R. 1937; Tropical Netherlands 1938; British Honduras 1939; Portugal 1941, 1958; Tanganyika 1942, 1948, 1956; (Belgian) Congo and Ruanda Urandi 1948; Denmark 1949; Belgium 1951; Australia (Atlas of Australian Resources) 1952; Poland 1953; Sierra Leone 1951; Morocco 1955; Israel 1957; India 1957; United States 1957; Kenya 1959; Arab World (published by Macmillan and perhaps not strictly a national atlas) 1960.

Apparently the Finns, though under the control of Russia at the time, can claim to have produced the first national atlas, compiled by the Geographical Society of Finland.[43] After a general topographical map, three physical maps deal with relief, geology and quaternary deposits; then follows a most interesting series of meteorological maps, including the monthly averages for 1881–90, isobars and wind roses, the number of days with temperatures over 0°, 5°, 10°, 15°, 20° C., the maximum thickness of the snow cover in four successive winters, the first and last frosts and the exceptional summer frosts. Later editions of this atlas included a comparable range of information which is obviously crucial in so northern a land; for example, the limits of various trees and cultivated plants, the distribution of state forests, and the types of forests are shown. On population, there is a detailed analysis of the statistics available for 1775, 1825, 1850, 1865, 1870, 1875, 1880 and 1890, with age and sex pyramids on which for all censuses except 1775 the rural and urban returns are separately shown. Current demographic information is classified by the death-rates, birth-rates, and marital status. This section, a series of diagrams on squares or squared paper, is followed by a map showing population density per square kilometre, and by a series of crops for the major local government divisions in the country; these include the proportion of people living in towns, those speaking Swedish, Finnish or other languages, having various religious affiliations, those blind, deaf-and-dumb or mentally afflicted, the illiterate and semi-literate, the percentage in the professions and of various classes of the population—the nobility, clergy, *bourgeois*, peasants and others. Mobility of the population is shown in a map of origin, which shows the percentage living in the commune where they were born, those living in the same province but in another commune, and those who have come from another province. Another social map shows the location of schools. Two maps deal with the harvest of 1893–5, with the cattle and butter sent from various parts: there are also flow diagrams showing the volume of exports and imports to each port. One map of special interest deals with the waterfalls, and the power used at each, and another with metals and rocks exploited. Communications are represented by three main maps, of which the first shows lighthouses and pilot stations on the inland waterways, the second railways, roads, minor roads, paths and canals, and the third the volume of passenger and goods traffic for 1895. Two practical maps deal with telephone and telegraph services, and postal routes, and the atlas finishes with archaeology and history. Clearly, this is an atlas of great interest and practical value; not all these maps were retained in later editions, but others were added including some fine general land-use maps which, with other data, made possible the regionalization of Finland discussed on pp. 137–40.

The 1906 *Canada Atlas*[44] was also of considerable practical importance at the time. It includes a statistical section for 1901, with a summary of the developments from 1867: there are also graphs of population growth for towns of over 7,000 people. The density and the origin of the major part of the population is shown only for administrative divisions. Basic maps include political units, relief, geology, minerals and forests, divided into three classes, northern, southern and Cordilleran, and further information is given on vegetation in the map showing the limits of forest trees, including prairies and the mixed wood and prairie country. As in the Finnish atlas, there are maps showing telegraph and telephone services, railways, inland and coastal waterways; others deal with mineral wealth, trade and commerce. Historical aspects are represented by maps showing the routes of explorers, and the location of some past-disputed boundaries. Subsequent atlases of Canada naturally show a far more sophisticated technique, but in so rapidly changing a country an atlas like that of 1906, which in effect shows conditions about the Census year 1901, is already a valuable historical document.

Both these early national atlases gave a great deal of basic information, but some of their modern successors show clearly the effects of long-continued research on particular problems. There is, for example, some fine work on physical features in the National Atlas of France[45] which is obviously due to considerable field observation; and in the *Atlas de Belgique*[46] an interesting physical map is given which shows the plains as coastal marsh, areas of glacifluvial and aeolian deposits, and river plains less than 50 m. high. Above this, from 50–150 m., 280–300 m., 380–400 m., 480–500 m., and slopes joining them, there are peneplanes of post-Hercynian and pre-Cretacean origin. Many local features are distinguished, such as fluvial terraces, the residue of meanders, sub-karstic plateaux of chalk sub-soil, escarpments due to rock resistances, monoclinal slopes, crests of an Appalachian type and sand-dunes. Another interesting map shows with particular delicacy the distribution of forests; and there are some interesting historical maps showing population developments in the country from 1846–80, 1880–1910, 1910–30, 1930–47. Although many modern national atlases have some very fine maps, not all have followed the pattern set by the Swedish geographer, Sten de Geer, who used dots to show population distribution (pp. 126, 185); for example the French national atlas, though having some superb maps showing the excellence of colour and the clarity of expression characteristic of a people combining taste and logic, is disappointing in its presentation of population distribution, for the compilers have succeeded only in showing the relative density by administrative districts.

The national atlas of Czechoslovakia[47] gives a fine series of maps on fifty-five double pages. Of these the first fifteen are given to physical geography, including a fine series on climate and vegetation; then follows a series of maps on population of special interest. The first shows the density per square kilometre on a scale ranging from 0–20 to 400–700, and then follows an informative group on the various nationalities, Czech, Russian and Ukrainian, German, Magyar, Polish, Romanian, Jewish, Yugoslav according to the 1930 Census. Statistical maps show that from 1910–30 there was an increase in the number of Czechs in part of the Sudeten territory, but a decrease of Germans from 1921–30, and also a decline in the number of Magyars in the area near the Danube. Ruthenia, handed to the Ukraine republic after the 1939–45 war, appears clearly as a separate element in the state. A wide range of social statistics includes religions, age and sex data, birth- and death-rates, the incidence of diseases causing deaths and of suicides, and both internal and overseas emigration. A soil-map acts as an introduction to a dozen maps dealing with various aspects of land use and agriculture, including farm sizes, and ten maps are given to industrial production, with one on electricity and power supplies. Two maps on transport show the use of waterways and railways, with the number of cars per 1,000 people, and even the use of telephones. Finally there are maps showing co-operative enterprises, holiday centres and spas, international trade, educational facilities of all descriptions, and social organizations, especially for sport and gymnastics. In short, the atlas gives a wide range of data on the physical environment and on the economic and social life of Czechoslovakia.

Here it has only been possible to give some indication of the type of material contained in some national atlases, which must vary from one country to another. It would be the most odious of comparisons to assess the pioneer atlases of Finland and Canada, noted above, against those produced within recent times. Modern methods of colour printing, combined with fine draughtsmanship, have produced splendid results, but the real merit of national atlases lies in their use of the research material available; for example, the French atlas has some fine maps showing geomorphological features, the Finnish Atlas has several informative maps on the distribution of farmland and forest with an interesting general regional scheme to which reference is made on pp. 137–40, and the Soviet Atlas includes some of the work on soils for which the Russians have long been known. Yet the purpose of such an atlas is not merely to exhibit the particular strength of research workers, but rather to exhibit the main geographical features of the country it presents. For this reason one turns to such physical maps as those of Sweden or Finland which show the glacial deposits and limits of marine and lacustrine beds, knowing that this gives some key to agriculture and rural settlement; it is appropriate that the French Atlas should contain some plans of its medieval towns, or that the Russian Atlas should show some results of the technological and industrial expansion of the present age. Although every national atlas will contain some standard maps of physical features, climate, vegetation, communications, minerals and the like, as, in fact, atlases have done for more than a hundred years, yet there is room for wide deviation from one to another in detail; it is also of value if some particularly interesting areas of each country can be represented—for example, certain major industrial concentrations or rural areas of special interest.

CHAPTER ELEVEN

NEITHER A BEGINNING NOR AN END

Geographers and their work; the attraction of geography;
some comments on geographical method.

A SURVEY of geography during one hundred years cannot be more than an outline, a presentation of episodes, a brief and perhaps at times tantalizing glimpse of personalities and their work. Inevitably, it has not been possible to cover the whole development of geography for a century, especially as each country has its own geography, varyingly studied it is true but probably in no country completely revealed, however great the output of its writers may be. Initially the idea of this book developed from the view that the modern academic pioneers of British geography had received inadequate attention in biographical literature; but though they are still studied less closely than their colleagues in America, the Institute of British Geographers now includes a bibliography with its obituaries, as the American Association has done for a considerable time. There are few biographies of geographers, though H. R. Mill and Griffith Taylor[1] have provided interesting autobiographies which reveal not only their intriguing if diverse personalities but also their clear conception of the work they should do in their time and place. Among other biographers, mention should be made of David Lowenthal's *George Perkins Marsh*,[2] not least because the author shows that many supposedly new ideas are by no means new: Marsh was the clear-sighted exponent of conservation, and saw that man made the earth rich or stripped it bare, according to the wisdom or folly of his choice. Even one generation of settlers in a new land such as New Zealand or western America may do irrevocable harm, for the relation between man and the earth is a delicate adjustment. Perhaps Shakespeare was right in saying

> 'There is a tide in the affairs of men
> That, taken at the flood, leads on to fortune;
> Omitted, all the voyage of our lives
> Is bound in shallows and in miseries.'

Geographers and Their Work

It is geography, rather than geographers, that provides the main interest, yet one cannot separate any man's work from his personality. The recent centenary of the deaths of von Humboldt and Ritter has led many people to consider again their views of geography, with varying assessments, including in some cases a marked dislike of Ritter's teleological religious approach.[3] As modern academic geography developed earlier in Germany than in other countries, the influence of its workers has been considerable: this has been shown particularly in R. Hartshorne's *Nature of Geography*.[4] Many ideas prevalent in British geography undoubtedly originated in Germany, but some were acquired indirectly through French sources. Before the 1914–18 war, many young British geographers went to the German universities as graduate students, and the journals of the time included a number of translated German articles, with summaries of many more. But from 1919, when British universities were acquiring honours schools of geography, the French influence was strong; students of the French geographical work of the time, however, will observe that though it possesses its own distinctive qualities—above all its clear and logical presentation—its writers were fully conscious of their debt to German scholars, and particularly to the great German tradition of atlas production.

Turning to America, no easy synthesis is possible. As shown on pp. 40–2, G. P. Marsh and A. H. Guyot were early pioneers, but with no successors, probably because Marsh spent much of his life outside America and Guyot left behind a number of books and papers that, though useful in classrooms and to those who wanted to know some meteorology and the heights of mountains, could hardly give anyone a thrill. A far more arresting figure came on the scene in W. M. Davis, a man of vast literary output who turned from meteorology to geomorphology and had much to say on the need for geography in modern education. Part of the initial impetus in his geomorphological work came from the pioneering work of men like G. K. Gilbert and J. W. Powell, who were the academically-minded explorers of the United States and possessed both the faculty for clear observation and the imagination to understand the relevance of what they saw. With his disciples, notably D. W. Johnson and C. A. Cotton, Davis made a strong contribution to geomorphology, even though many of his concepts,

247

such as the cycle of erosion, are now subject to criticism which at times becomes total denigration. It now appears that an analysis of slopes is a prime need in physical geography, and one admires the enthusiasm of a geographer at the 1960 Congress in Stockholm who wants—doubtless with assistance—to measure every slope in the world. Davis was in close touch with distinguished European geomorphologists, notably Penck of Berlin; and as this study grew in America, so it advanced in Europe, where some of the most famed geographers, such as Cvijić of Belgrade and de Martonne of Paris became regional workers but also made substantial contributions to geomorphology.

In spite of the impetus given to geomorphology by W. M. Davis, modern American geography has become primarily a human study in the widest sense, with the physical aspects taught mainly in schools of geology. Among many notable figures, Isaiah Bowman stands out for his great contribution to political geography in his *New World*, which helped many students of the inter-war period to understand the complex and problem-ridden world in which they found themselves. Like Cvijić and de Martonne, Bowman was initially a physical geographer, but in maturity his mind turned to a wide range of human problems, including especially those of population distribution. But of all enterprises in American geography, one of the most fruitful was the excursion of a few young men who before the 1914–18 war went out to see what they could observe of human geography in the field.[5] In various ways, through surveys of land-use, agriculture, settlements, towns, they and others made human geography into a field study. The relevance of field study is very clear in geomorphology, but not always as apparent in work on the human side; in America much of the work on economic and social aspects acquired strength from the study of the living landscape. At the same time, one must recognize that statistical and other material is contributory to geographical study, and not unusually can be mapped with informative effect.

Actual study of the landscape is generally regarded as essential, but the power of discriminating observation differs widely from one person to another. The Americans who pioneered in field-work to satisfy their interests in social and economic geography were only doing what Frenchmen, Germans and others had done before them; on p. 89 it was noted that Cvijić wrote his great book on the Balkans primarily by walking through the hills and valleys he knew so well. Equally, Vidal de la Blache's *Tableau de la Géographie de la France* breathes the spirit of the landscape: one feels that the author has not only been there, but is anxious to make the reader go there too in imagination (see p. 84). Now, fifty years and more after these inspired writers walked through their country-sides, modern geographers proceed with far more elaborate schemes of land-use mapping, of recording the varied uses of buildings in villages, or of interviewing farmers. Supplementary evidence is more abundant also, and a notable recent gain has been aerial photographs. In town study also, the technique has become more elaborate, and various forms of mapping are supplemented by the use of statistics and, in the hands of many workers, by interviewing. The method may be more elaborate; the purpose remains the same.

Academics not uncommonly group themselves into various schools of thought and colour their work by their general views on life as a whole. Complete detachment is probably unattainable; certainly if one looks at the progress of other branches of learning, it is not invariably found. Economists may reveal conservative or Marxist views, and Macaulay is generally regarded as the great Whig historian of his time. In recent years, some continental geographers, notably in France and—naturally enough—in Russia, write with clear Communist sympathies, and the arguments of the German geographers for an expansion of the national territory, though strong during the Nazi period, date back at least to the beginning of the present century. In many cases, political views became nationalistic, and the study of geopolitics, though generally regarded with horror by geographers, was really an effort to set an academic seal on theories of nationalism in relation to territory and to make what seemed to some to be a more logical arrangement of political boundaries.[6] The attack on geopolitical thinking rests on the claim of its exponents to have established a new science, based on a combination of geographical factors with political aspects and some form of sociological reasoning. So much that has happened in geography had had some motive force of political origin: look, for example, at the impetus given to French geography by the colonialism of the period from 1871 and the frankly-expressed policy of some French geographical societies of encouraging the acquisition of new territories in Africa and elsewhere.

In the modern growth of geography, no time was more vital than the years 1910–20. The lights sank down on the Europe politically designed in 1815, and out of the night of war there came the hope of a new and more peaceful day for a continent that, within a hundred years, had acquired a vast increase of population, new and powerful industries, wide colonial responsibilities and an international trading system that, though disrupted, must inevitably be revived if millions were to be given even the first essentials of decent living. The word 'reconstruction' was in the air and the idealist of a former time became from force of circumstances the realist, the practical man of the moment. In such an environment of thought, men like Cvijić of Belgrade and Romer from Poland found that their work on their national territories gave a basis for countries on the map of Europe. As noted on p. 220, it is not clear to what extent geographers helped to make the new map of Europe, along with historians, economists and others, but at least their fundamental contribution of distribution maps, and the cartographical presentation of possible frontiers, was respected at the Versailles sessions: eventually, more of the story may be told when the papers of Isaiah Bowman are unsealed. Reconstruction was much more the drawing of new political frontiers and steadily ideas of replanning towns and country-sides became fashionable; in Britain, Patrick Geddes and a few others found that at last there was some hope of the effective reorganization of cities, and within a few years vast new housing areas had been added round the towns of Britain. It is a widespread fallacy that planning is a new idea: apart from the fact that it probably goes back to the earliest days of living in cities, and that the Romans were keen planners, Britain has several cities that show a clear imprint of Georgian planning. To some extent planning was a casualty of the Industrial Revolution, which in some countries came with such speed that long-abiding mistakes were made in the provision of inadequate houses, unsuitable town centres and badly located industries.

Planners are often thought of as a group quite distinct and apart from geographers, but in all planning the essential rule was given by Patrick Geddes, 'survey before action'. Had the regional survey of Britain, and doubtless of other countries, been more adequate, and had there been more study of both towns and rural landscapes, then many of the plans could have been made with greater ease and probably greater effect. In the Land Use survey of Britain carried out during the 1930's when agriculture was a depressed industry, a foundation was provided for a measure of regeneration which is readily perceptible in the changed landscape of rural Britain in

1960 compared with that of 1930. The need now, being courageously met, is for a resurvey of agricultural Britain, and for a Land Use survey covering the whole world.[7] In town study, there is a need not only for those fundamentally economic and social enterprises that consider the relations of towns to areas around them, but also for detailed analyses of particular towns, of which some distinguished examples are noted on pp. 193 and 198. In the aggregate, as a recent bibliography shows,[8] such work is considerable, yet when one begins to search the literature, very little will be found on many towns and cities of considerable size. There is, in fact, no standard technique, for French, German, British, American and other geographers have attacked the problem of town study from many angles; in the opinion of the present author —an opinion with which many will disagree—it is desirable that a variety of methods should be tried at present. Put it bluntly—a great opportunity has been seized only partially.

Another challenge to geographers lies in the rapidity of change under modern world conditions. In Britain this has been seen most notably in the outward spread of housing since 1919, and in the changes in industrial location by which some areas have declined markedly, particularly the older textile—manufacturing and coal-mining areas, and others have expanded, notably the areas around London and in the west Midlands. And in the north-east of the United States, the even vaster spread of townspeople into rural areas is now receiving attention through a research organization known as Megalopolis, under Jean Gottman.[9] There is a clear need both for an initial survey and for a periodical resurvey of any area: where this has been done, the results have been rewarding, as for example in Stockholm, where the papers of W. William-Olsson,[10] which appeared respectively in 1940 and 1960, show how great the changes have been—a fact which is apparent even on the most cursory visit to the city. And while it is desirable to watch developments as they come, some of the literature on cities has been vividly concerned with the past. Many French studies, including some pioneer efforts,[11] have adopted the not unusual French practice of taking a view of history from the beginnings of civilized occupation of the site; a number of German writers have been concerned particularly with the form—or, as some put it, the morphology—of towns.[12] Both these have a strong historical element, and so too have various studies which use old maps to find the actual area covered in the past by towns, and even to map their past as well as their present distribution of population.

From the present, therefore, it is natural to look to the past, particularly in countries where towns have a continuous existence of as much as a thousand years. It is not, however, quantity of history that matters, but quality, and one of the most interesting of all studies in historical geography is the *Mirror for Americans*[13] by R. H. Brown† (1898–1948), on the eastern seaboard in 1810 through the eyes of an imaginary author, T. P. Keystone. The rapid growth of towns in America has provided rich opportunities for town study and, as shown on pp. 194 and 198 some valuable work has been done; one might wish, however, that some authors were more conscious of towns as dynamic rather than static, allowing that they may decay as well as grow.[14] It may be that changes are more obvious in towns than in the countryside, yet rural areas change considerably, even within a period of a few years. Though in Ireland many like to dwell on the thought of the immemorial past, and though like Britain it has been settled for thousands of years, yet the present agricultural landscape shows very clearly the effects of a population decline which has had various associated effects: farms have been enlarged, old houses and field boundaries abandoned, estates of the landed gentry divided into farms, peat bogs partially or wholly cleared, and new houses added through private enterprise or with governmental aid. The vast decline of population, in many areas by two-thirds or even more since 1845, has left its clear imprint on the landscape.

The reconstruction of a past landscape is a fascinating art and may involve the use of documentary, map and field evidence, some of which may be hidden: in Britain, for example, M. Beresford has shown that lost medieval villages, of which some probably perished at the time of the Black Death, may be revealed by digging.[15] Aerial photography may be helpful if it shows lines not normally visible on the ground: in Britain, its value has long been recognized by field archaeologists and notably by O. G. S. Crawford.[16] Though it is generally recognized that the long-settled lands of the world, such as western Europe, China, and India, have landscapes made by man, it has not proved easy to reconstruct past landscapes in spite of some excellent work. The Ordnance Survey maps of Roman Britain have included estimates of the forest cover of the times, and so have followed a still older quest of relating the distribution of people to the intrinsic natural conditions: J. R. Green's *Making of England*, first published in 1881, opens with a map that shows forests, marshes, heaths, and mountainous country, and several regional maps of the same type have also been included. One purpose of historical geography, stated by E. W. Gilbert[17] in 1932, is to give a picture of the regional geography of the past, and an interesting volume of essays for England and Wales appeared in 1936 along these lines.[18] Since then, a series of volumes has been planned, of which the first have appeared, on the geography of the Domesday period by H. C. Darby and his collaborators.[19]

There is a clear attraction in studies tied both to some definite past time and place, and geographical journals include numerous examples of local work along these lines. Much can be observed in the field, but by no means all, and a vast amount of data, so far only partially investigated, awaits the researcher in map libraries, old government reports, estate plans and other sources. Especially is this true for the nineteenth-century landscape, which in many countries is vanishing quickly through the rebuilding of towns, the outward spread of housing and the modernization of agriculture. In some areas of Britain, this is far more apparent than others: for example, in much of the inner Pennine area of Lancashire and Yorkshire the textile towns have changed comparatively little for some time as the population has been stationary or declining, and the use of stone has given durability to the houses, mills and public buildings. True, coal-mining has ceased in the areas first worked, and the major pits are now penetrating deeper beds; but mining is by its very nature ephemeral. Maps showing the past compared with the present distribution of mining are commonly and rightly given in textbooks: such maps are a contribution to historical geography.

Various materials are being laid up for posterity in current geographical writing. Any regional geography, or study of its economic and social aspects, which faithfully shows what now exists, may prove to be of value at some future time; had some of the pioneer geographers of the 1870's been concerned to show what they saw near home, much that they might have written would be of value now. Instead, one turns to such sources as the journals of statistical societies of which some, such as that of Manchester, were then also deeply concerned with social inquiries and under this head left accounts of the use of particular streets for housing and commercial purposes, with excellently vivid discriptions of the living conditions of the time. In our own day, agencies such as societies and public libraries have combined to provide photographic records of towns, including in some cases every building in certain streets:[20] some of this may prove to be valuable source material of the future. But it is still part of most geographers' training to carry out some field-work, and just as in Britain some of the original work for the Land Use Survey of the 1930's was done by students, so it is expected that the more ambitious scheme of thirty years later will be achieved partly by student-workers (p. 169).

Whether or not geographical study in the past hundred years has been as fruitful as it might have been, one will not presume to say. But it could not have reached even its present stage of development except by attracting the interest of a vast number of people, both in schools and universities and—equally if not indeed more important—among the general public. For a long time it was complained that in Britain few young geographers were given the opportunity of working in colonial territories, or of acquiring posts in which their knowledge could be effectively used, at least as a basis for a life work. Much of this has changed, and at present many young British geographers are finding appropriate employment abroad, or at home, in planning departments, as well as in such central ministries as Housing and Local Government. In the United States, Canada, the U.S.S.R. and other countries, the opportunities for young geographers are considerably greater; one may note in passing that some geographers, notably Bowman in America and de Martonne in France, became national and even international figures. The entertaining autobiography of Griffith Taylor[21] shows that his comments on the aridity of central Australia caused certain newspapers to heap abuse on him, and that one of his books was banned for twenty years in Western Australia. Finally, however, he points out that everyone saw that he was right, and he ends with a quotation from Haldane that 'the characteristic part of any man's teaching is what is *novel and heretical* in it, not what he and his audience take for granted'. Incidentally, his statement that it may take twenty-five to thirty years for a new idea, or a new evaluation of the basic facts, to reach the textbooks, is borne out by the practice of one examining board which estimates that it may take twenty-five years to remove a fallacy from a school text. But not every heresy proves to be true in the end.

In this book, many geographers have appeared for at least a

moment on the stage, and then disappeared; others have appeared several times like members of a stage army; of some a brief biography—as much as would go on one index card—is given on pp. 303–25, with a note of biographical sources. It is not claimed that this section includes all the greatest geographers of the past century, though it includes many whose work has proved of interest; for obvious reasons no living geographer has been included. The range of geographical experience, and the variety of its expression in writing, is so wide that no student should lean heavily on the views of one man, as indeed many students do and, one must add with regret, are even encouraged to do; equally, to the author at least it is apparent that many ideas and new techniques are not new, but merely taken up again after an interval of years and more effectively developed. The attitude of geographers to others can be a most interesting psychological study: there is the 'great-men-in-other-days' school and the 'now-we-can-show-them-how-to-do-it' type, epitomized in the young research student who told the author that his thesis would set geography on its feet. The real difficulty lies in doing the work, and there can be no doubt that the advance of geography in the past hundred years is due to the writers of articles and books, and the compilers of maps, rather than to those who have merely taught and criticized or—worse still—merely 'administered'.

The Attraction of Geography

Viewing the past hundred years, the initial attraction was clearly exploration. And it has proved an abiding attraction, for the appeal of strange and rare circumstances and experiences still survives. In our times there may be few areas of the world entirely unknown, but audiences may still be attracted by visitors who have been to China or some remote part of the Soviet Union. The modern rise of geography has coincided with the great epoch of exploration, followed by the growth of trade and the founding of colonies. Geography had its due relevance to the merchants and industrialists of Victorian England, to the imperialist German Empire from 1870 and to the defeated France of the same period; in Australia and in the United States and other American countries, the interest lay in discovering and working out the potentialities of the home territory. By the 1890's there were signs that a group of academically minded Frenchmen were becoming tired of the perpetual *nouvelles à sensation*, and so they founded the great

Annales de Geographie, a journal respected everywhere. Though a reaction must come eventually, there are still large numbers of people who regard geography as primarily exploration of remote places; but in Africa, Australia, New Zealand and many other countries that were the subject of exciting revelations from explorers a century ago, there are now university and other geographers working with well-tried research methods.

Exploration was only a beginning, a reconnaissance. The next stage was to turn the results of explorers' journeys into a systematic account of a country, aided by the steady recording of observations on maps. After the main period of exploration was over, by the end of the nineteenth century, a natural development was the compilation of the 1:1,000,000 map for the whole world, to which one valuable contribution was the series of sheets of Hispanic America compiled under the care of the American Geographical Society.[22] At present there are schemes for recording population densities over the whole world on the 1:1,000,000 scale, and so providing maps of reasonable accuracy for many areas where the maps of population density are merely guesswork.[23] Much of the world is inadequately covered regionally: of China and its surrounding territories, for example, little was known from the days of Marco Polo, until A. Little used the travels of Sven Hedin to write *The Far East*, other travellers gave valuable supplementary information, and a notable effort was the Christian Occupation of China volume. But the cover of information is so inadequate that any regional geography of China must be based partly on guesswork—or at least on generalization from inadequate evidence; under such conditions local studies are useful as 'samples'. Aerial photographs may be a useful source of supplementary information but the real need is detailed field-by-field and street-by-street investigation.

World schemes of regionalization have given millions of people some idea of the whole globe, even though they may seem appropriate only in the earlier phases of geographical study. Just as the first developments of geography among the Greeks and others were due to a desire to know what lay beyond the immediate surroundings, so today some purpose is served by acquainting people with the general form of the world; schemes of world regionalization based on landforms and structure, climate, vegetation, land-use and population, are still of value. Virtually all school atlases include some maps of structural regions, based on a number of workers' classifications, including those of Suess and de Martonne. A refinement of such maps is seen in such maps as that of the geomorphology of Europe, by D. L. Linton[24] in Bartholomew's Advanced Atlas; somewhat significantly the geological map, by periods, is relegated to an inset. Another physical map commonly included is that of earthquake zones and active volcanoes, and some atlases are now including maps of soil types; there has been a vast increase in the knowledge of soils during the last fifty years, and in Russia, for example, soil types are sometimes used as a basis for regionalization of the vast lowland, in association with land-use.

Regionalization on a basis of climate has long been popular, though as the study of climate becomes more complex with the study of the effectiveness of rainfall in relation to its incidence, relation to evaporation and holding power of soils. The Herbertson classification of 'natural regions' on a basis of amounts of rainfall and critical temperatures ($0°$, $10°$, $20°$ C.) goes back to German sources and notably to Köppen; but it now seems inadequate and naïve as modern workers have shown how complex climatic study can be. Ultimately some new classification of climatic types may appear, some new synthesis from the present phase of analysis. Nevertheless, schemes such as those of Herbertson have served, and indeed still serve, a most useful purpose in giving a world view. For Europe, the division into Mediterranean, west maritime, various grades of continental climate, the tundra fringe, and mountain climates, gives some foundation for further study; such a division bears some relation to the types of vegetation, the crops, the seasonal rhythm of agriculture, and even to the construction of houses to meet particular conditions of weather and climate. Having steadily acquired the basic data on climate, inadequate as this may be, in records of a number of stations, the need now is to refine techniques by further investigations, and two main methods are followed. These are first, the close study of small areas, or microclimatology, for here as in other aspects of distributional investigation, the discovery of what happens in a small area may be highly indicative—an early example is given on p. 304 for the English Lake District; the second valuable detailed study is of weather, particularly in areas where, as in Britain and Ireland, polar, maritime and continental influences are all interwoven. An example of the value of careful observation of weather in relation to glaciers is given on pp. 112–15. Summarizing the argument of this paragraph, it is clear that world classifications are only a beginning, and that though such generalizations are based on local data, they lead workers back to the accumulation of more local data.

World schemes of vegetation have been produced for over a hundred years, and the recognition of such associations as those of savanna, steppe, selva, has become general. Probably many readers of this book remember the pleasantly coloured wall maps of Marcel Hardy, originally a pupil of Patrick Geddes, and have also studied his textbooks.[25] Biogeography, though now somewhat neglected, was a popular study sixty or seventy years ago, partly because it revealed the relationship of vegetation and climate: the idea of the interdependence of all physical phenomena, of the unity of the physical world, inherent in Darwinism, was regarded with esteem. Reconstruction of the natural vegetation, however, was difficult in much of the world as the human imprint was so strong that few if any examples of a natural landscape survived. Although the idea bore fruit, in time more attention was given to the actual land-use, for in such anciently cultivated areas as China or western Europe hardly any trace of the natural vegetation might remain; indeed many fallacies were put forward, such as the view that large areas of the English chalk country were in a natural state grasslands, though in all probability they had been woods, not of necessity continuous, and the supposed natural grasslands were, in fact, due to the nibbling of sheep which kept back the growth of trees and shrubs.

Population distribution over the world is crucial, and the real fascination lies in its inequalities. At present, as noted on p. 190, there are hopes of a far more adequate treatment of population distribution than at any previous time, but the existing maps, though generalizations on inadequate data, at least give some indication of the interesting inequalities between one area and another. The vivid work of Vidal de la Blache[26] on the areas showing marked concentrations of population is of great interest, and the emphasis of many workers on excessively high densities in certain countries has proved stimulating; partly from the revelations of density maps many have been led to consider such problems as overpopulation and—its inevitable counterpart—underpopulation, and to ask if there is, or can be, an optimum density of population. Such issues were crucial in the days of heavy unemployment of the 1930's, but the problem of high densities of population in relation to natural resources remains a permanent one. Vidal de la Blache and others gave a world view, and though that can provide some framework for general thinking on the subject of food and people, the real fascination of population study lies in its detail. A great advance in technique came with the work of Sten de Geer (p. 185) in Sweden, which has been widely emulated elsewhere, but in the present author's opinion not widely enough, as many 'standard' maps of population density have a far lower degree of accuracy than the Swedish method provides, highly laborious as it is. Mapping of population densities on a basis of administrative areas is a quick job, but hardly an efficient one, as within even a small administrative district, the population may be most unevenly distributed. When one is served with maps of population distributions of areas such as an English county, the result may be ridiculous.

All the above distributions may be, and have been, used as part of a regional synthesis. Throughout the preceding few pages, it has been implied that the world generalization has been useful and enlightening, provided it is not regarded as anything more than an approximation to the truth: the real need is to get back to the local detail. Without the climatic regions, the Marcel Hardy vegetation maps, the world population maps, geographical education could not have advanced to its present level. Equally one must admire the vivid physical maps such as those of some great German firms, yet they have some of the dramatic quality of a Wagner opera— fascinating, colourful, arresting, but quite definitely larger than lifesize. On a simple scale, the correlation between physical features, soils, climate, vegetation, land-use and population distribution may be made quite easily; and it was this unity of man and the earth that provided the attraction of geography. Indeed, it still is, if one follows R. Hartshorne's view[27] that 'the unique purpose of geography is to seek comprehension of the variable character of areas in terms of all the interrelated features which together form that variable character', or again, 'geography is concerned to provide accurate, orderly, and rational description and interpretation of the variable character of the earth surface'.* The appeal remains the same—observed variety and change from one area to another which challenges analysis and understanding. If one investigates even a small area, of a few square miles, there may be differences

* This definition, however, is twice modified, first (p. 47) 'to describe and interpret the variable character from place to place of the earth as the world of man', and (p. 172) 'the study that seeks to provide scientific description of the earth as the world of man'.

of scene that offer no immediate explanation, but may prove to be due to variations in soil and drainage conditions, to the varying economic history which may have its mark in relics of past mining and quarrying (still conspicuous in many upland areas of Britain), or to widely-ranging standards of efficiency between one farm household and another. There are areas of the world where for many hundreds of square miles there is little change of scene: the author has not had the experience of travelling by train across the Australian desert, but he is informed that for hundreds of miles there is no apparent change of scene, though possibly the trained expert on desert landscapes would see much of interest. But those whose geographical training and travel was mainly in western Europe will know that the whole challenge of their fieldwork lies in the swift changes of scene, due not only to fundamentally geographical but also to historical causes. The Hartshorne definition involves much more than might at first appear.

Some Comments on Geographical Method

Thirty to forty years ago, many geographers were attracted to the idea of broad generalization and their students were given the stimulus of facing large ideas in university courses. And there are still teachers working with such aims, though many world syntheses have been challenged by the advance of detailed local regional work; the one-time human courses which considered the geographical influence on history from the beginning of time to the present are probably fewer in number than formerly. No doubt there are still political courses which attempt to survey the modern scene with the sweep of Bowman in his *New World*, and economic courses of comparable scope. But the general trend has been towards more specialization, both by the defining of regional courses within more restricted limits, and by the concentration on particular topics within the field of economic and social geography. This development has been made possible by the vast addition of material within the past thirty years; although there are many gaps, and many 'long-felt wants', yet the production of books and paper is considerable. The trend towards specialization perhaps reaches its extreme expression in America, but it is also apparent in France: deplored by some, it has the aim of dealing thoroughly with some particular theme. What has been called 'topical' geography—not in the sense of dealing necessarily with the current time but rather with a topic—may, indeed should, lead a worker back to the fundamental issue of the earth and man. If for example a writer deals geographically with electricity in a country, he will be led to consider the home resources of fuel, including water, and those imported, the ease of transmission, the form and distribution of power stations, the ultimate destination of the supply, and probably the location of industry. He may come to the conclusion that a state uses a home-fuel at greater expense than one imported for reasons of national economic and strategic policy, and for the servicing of industries established partly to give employment in remote locations. The modern provision of electricity, has, in fact, made possible a redistribution of industry throughout the world. With electricity some European countries, such as Norway, Sweden and Switzerland have advanced markedly, and in older industrial countries such as Britain power transmission has opened many new areas for factory development, conspicuously the London area and the West Midlands. An apparently simple theme, like an inquiry into electricity, may lead to a wide range of investigations.

In regional study, the tendency of some writers is to base their work round a theme, such as the distribution of population, and from that to consider the relation of the people to the earth, by considering the effects of physical features, climate, soils and other features on the agriculture which supports some, at least, of the population. This is a departure from the stricter orthodoxy of regional treatment, which proceeds from the physical features to those of human origin; indeed some American geographers have been troubled by the development of a 'dualism', or rigid separation between physical and human (or 'natural' and 'cultural' in American) geography, combined with the virtual elimination from the geographical departments of all geomorphology. This is a departure from the views of the earlier geographers and to the present author it seems that a study of the physical environment can only enrich the general treatment. If, for example, one studies a town, careful observation and analysis may reveal that its form and layout bears an interesting relation to physical features, with the wealthier suburbs in the more desirable locations on higher ground, or by the sea, according to local circumstances; equally significant, in many towns the centres have remained on much the same site for hundreds of years, even possibly more than a thousand years, as in Chester where the Romans occupied an outcrop of sandstone which gave a dry site near to a possible crossing-place of the River Dee. Town study has been mentioned, as perhaps in such work the relation of physical features to the present form may be obscure; undeniably in some areas, it is. Given a virtually level site, such as that of south Manchester, the designer of streets may work on his drawing-board and provide a rectangular pattern without disaster, but the effect may be most unfortunate if he uses the same design over hill over dale as in some American cities, or on steep-sided drumlins as in Glasgow.

There is every reason to guard the idea of relating any distributions to fundamental aspects of physical geography; in the primary settlement of colonists, for example in Australia, a knowledge of physical geography could have saved people from considerable hardship. Recently E. G. Bowen has shown that the Welsh colonists in Patagonia were initially misled by fallacious accounts of the climate and soils, but that after a time they moved their settlements from the coast to the interior with fortunate results.[28] Circumstances are widely different from one area to another. In Britain, town expansion is a problem because so many of the desired sites prove to be on land of unexpectedly good quality or scenic amenity; recently in touring the new suburbs of a rapidly-expanding Stockholm, the author noted that almost all are on rock outcrops that, like much of the Baltic shield, are covered with very little soil, valueless for agriculture and supports only a poor type of forest. How happy the planners of Stockholm should be to possess such sites so close at hand! On the use of sites for new housing, opinions differ widely. The Swedes have chosen to mark the centres of their new suburbs by erecting flats of some twenty storeys, partly to ensure that large numbers of people live near the shopping and social centre, but also to give emphasis to the *centrum*; near Manchester, however, on an overspill estate in Cheshire, protests have been made against a proposal to erect buildings of some ten storeys as they are held to be an alien element in the rolling Cheshire countryside and an infringement of the view of the Pennine Chain a few miles away. In the Swedish example, the effect is a frankly urban landscape; in the British example, the desire is apparently to make the urban element inconspicuous on the argument that the rural landscape must be maintained to the maximum degree possible.

A matter of some controversy is the extent to which a geographer's work should give directive force to such developments as the replanning of cities, the improvement of agriculture, the provision of suitable transport facilities, the relocation of industries. Any geographer who, like the present author, has written on matters which raise social problems, is likely to find his work reviewed by earnest young men who criticize him for not advocating green belts round cities, for not making a plea for new towns, in short for advocating or not advocating this or that, according to the predilections of the reviewer. One must, it seems, have a solution to offer, a doctrine to preach. It could be argued that pure geographical inquiry, as such, has no solution to offer for any problem, but that it can provide the material for an intelligent assessment of certain problems and in so doing contribute to their solution. At this stage the geographer applies his findings to a set of circumstances, and may suggest some change, some development. A notable example of a pure geographical inquiry in Britain which became of practical significance was the Land Use Survey: initially a fact-finding enterprise, it became basic to the agricultural renovation of Britain which was initiated during the 1939–45 war and has continued ever since. Had the ground-nuts scheme in East Africa been preceded by a sound fact-finding inquiry, many millions of pounds would have been saved.

All this raises questions of the place of the geographer in planning. At least one distinguished British planner has said that planning is essentially applied geography, and geographical inquiry is the first step towards reconstruction. In Britain, for example, surveys of derelict land, of which that by S. H. Beaver[29] in the Black Country was a pioneer example, have been the prelude to a considerable use of such areas for houses, industrial sites, parks and playing-fields; without the use of such land the problem of rehousing many thousands of people in the Black Country would have been more acute. Equally, derelict land has provided factory sites in South Wales valleys, where level land for industry is scarce. One appears to have come back to the Geddes watchword, 'survey before action'. The planner, however, needs much more than a geographical training, for in his work he may require to know something of law, of architecture, of engineering, of economics—in fact of almost anything; as a basic training, geography is valuable but its application will raise many problems, not least those of a political and financial nature. Nevertheless, had there been a more thorough geographical survey of Britain by 1939, some valuable material would have been ready for the planners who, in many places, were obliged to conduct rapid surveys under difficult conditions.

In many and varied ways, the ability of geographers to do a practical job has been demonstrated, but some may perhaps look back with interest, and a few (not the author) with nostalgia to the days when geographers appeared to have the key to the explanation of civilization. All world strategy, it seemed, was explicable in terms of Mackinder's Heartland and its fringes; Huntington saw climate, and its pulsations, as the key to human history and civilization; and Ellen Churchill Semple brought forward stimulating ideas on environmental determinism which were followed enthusiastically by many others. Reaction set in, particularly among the French possibilists who so viewed the changes and chances of this life that they could not accept the view that there were four or five great influences that moulded humanity. To many, the idea of a general law is most attractive: Griffith Taylor,[30] for example, said that 'nature's plan was obvious': man had only to study 'the character of the environment' so that he could follow the plan 'determined by nature'. But who is to decide what shall be done? Those who have most thoroughly studied the character of the environment? And what if they disagree? In any case there may not be one most suitable use for a particular area; there may be conflict between forestry or sheep-rearing in mountains, between forestry or farming in lowlands, between dairying or crop production on good agricultural land, between preservation for agricultural use or allocation to town overspill, between mineral exploitation involving temporary—even permanent—ruin of an area for agriculture or its continued farming. Nor can choices be made solely on the study of the 'character of the land', for national policies may have an important, even a directive influence, on the use of land. It is Dutch policy to acquire new land by expensive engineering works, British policy to safeguard land as far as possible against urban expansion, American to allow a far freer use of land for commuters than Britain, Belgian to allow, even encourage, the provision of smallholdings to be worked by townspeople; each country has its own needs and traditions. But while there may be no one single course of right action determined by nature, it is also very clear that by unwise and often optimistic expansion into new areas, especially in the United States, soil erosion has been made a deadly scourge, by the removal of trees virtual deserts have been constructed, and by mining with no provision for the dangers of subsidence or land restoration thousands of acres have been wasted or reduced substantially in value.

Against such a background, the issue of determinism and possibilism as mutually exclusive theories may seem unreal.

Through this book an attempt has been made to show some of the varied work of geographers during the past hundred years. The story opened in the working years of Ritter and von Humboldt, and continued with the great phase of modern exploration, and an allied if somewhat halting and uncertain educational expansion. Exploration opened up the world, and in time various writers arranged and synthesised the often formless mass of data that travellers provided. As cartography advanced, both through government surveys and the activity of a number of excellent private firms and individual workers, the globe and the atlas brought knowledge of other lands to everyone who sought it—indeed a globe was quite commonly part of the furniture of the intellectual Victorian home. Universal geographies became available in the latter part of the nineteenth century, to be replaced by others in the twentieth, as well as by steadily growing series of textbooks. Material was culled from many sources to establish some general principles, such as those of human geography, or some of the regional schemes of presentation; the French, following a tradition going back to the early nineteenth century but apparently laid aside for a long time, finally gave particular distinction to the study of regional geography. Drawing some inspiration from the allied study of geology, various writers such as Suess, Penck and W. M. Davis strengthened the interest in geomorphology, and from a variety of sources there came data on climatology, of which in Britain H. R. Mill was a distinguished exponent. Many curious developments included the growth of geopolitics on the continent, particularly during the Nazi period in Germany; a fascinating speculation, on which more evidence is wanted, is the exact influence of geographers on the Treaty of Versailles and subsequent settlements. After the 1914–18 war, many geographers in Britain were clearly concerned with the prospects of replanning Britain (in fact a halting enterprise until after the 1939–45 war) and of establishing world peace. Since 1945 there has been a further expansion in university education, and new techniques of research are being tried with good effect: the swiftness of changes in the present world makes the challenge to the researcher at least as great as at any previous time.

NOTES AND REFERENCES

THIS book deals only with a recent phase of the history of geography, and those looking for material on earlier times should consult the three classic works listed under Ref. 1, Chapter One and Ref. 10, Chapter Two. Two excellent recent assessments of French and American work are listed under Ref. 6, Chapter One. The most comprehensive study of geographical methodology is Hartshorne, R., *The Nature of Geography* Annals of the Association of American Geographers, 29, 1939, republished in 1946, 1949, and 1959: a recent commentary on this work, and all the controversy it has evoked, is the same author's *Perspective on the Nature of Geography*, 1959. Griffith Taylor's *Geography in the Twentieth Century*, 1951, also 1953 and 1957, is a compilative work. There are many other books and papers on the content and method of geography, and a substantial bibliography is given in the two Hartshorne works, which is particularly rich in German references. Of works dealing with societies, perhaps the most comprehensive is that of J. K. Wright on the American Geographical Society, Ref. 15 in Chapter Three, but the small volume, by H. R. Mill, on the Royal Geographical Society, Ref. 5 in the same chapter, is inadequately known. Biographies of individuals are comparatively few and many obituaries are regrettably obscure as source materials, but David Lowenthal's study of George Perkins Marsh, 1958, is a notable full length work and there are some interesting articles in journals on past geographers. Few geographers appear to write their memoirs, but two interesting exceptions are H. R. Mill and Griffith Taylor.

In the references that follow certain abbreviations have been used for journals and series used several times. These follow the list given in the World List of Scientific Periodicals:

Advanc. Sci. Advancement of Science, London.

Ann. Ass. Amer. Geogr. Annals of the Association of American Geographers, Minneapolis.

Ann. Géogr. Annales de Géographie, Paris.

Biogr. Mem. Nat. Acad. Sci. Biographical Memoirs, National Academy of Sciences, Washington.

Bull. Amer. geogr. Soc. Bulletin of the American Geographical Society of New York.

Bull. Amer. geogr. stat. Soc. Bulletin of the American Geographical and Statistical Society. New York.

D.A.B. Dictionary of American Biography.

D.N.B. Dictionary of National Biography, London.

Econ. Geogr. Economic Geography, Madison, Wis.

Geogr. J. Geographical Journal, London.

Geogr. Rev. Geographical Review, New York.

Geogr. Teach. Geographical Teacher, London, later *Geography*.

Géogr. Univ. Géographie universelle, Paris.

J. Amer. Geogr. Soc. Journal of the American Geographical Society 1871–1900, New York.

J. Amer. Geogr. and Stat. Soc. Journal of the American Geographical and Statistical Society 1859–70.

J. Geomorph. Journal of Geomorphology, New York.

J. Manch. Geogr. Soc. Journal of the Manchester Geographical Society.

J. R. Geogr. Soc. Journal of the Royal Geographical Society, London.

Petermanns Mitt. Petermanns, A., Mitteilungen aus J. Perthes's Geographischer Anstalt, Gotha 1885–1938. The journal is now Petermann's geographische Mitteilungen.

Proc. R. Geogr. Soc. Proceedings of the Royal Geographical Society, London.

Pub. Inst. Brit. Geog. Publications of the Institute of British Geographers, London.

Scot. Geogr. Mag. Scottish Geographical Magazine, Edinburgh.

Supp. Papers, R. G. S. Royal Geographical Society, Supplementary Papers.

CHAPTER ONE

1. BEAZLEY, C. R., *The Dawn of Modern Geography*, 3 vols. Vol. 1 (to A.D. 900), London 1897; Vol. 2 (900–1260), London 1901; Vol. 3 (1260–1420), Oxford 1906: BUNBURY, E. H., *A History of Ancient Geography*, 2 vols., London, 1879: TOZER, H. F., *A History of Ancient Geography*, Cambridge, 1897, 1935 (with notes by M. Cary).

2. DARWIN, SIR CHARLES, *The Problems of World Population*, Rede lecture, Cambridge, 1958, 37.

3. MACKINDER, H. J., 'The Progress of Geography in the Field and in the Study During the Reign of His Majesty—King George the Fifth,' *Geogr. J.*, **86**, 1935, 1–16.

4. VAN CLEEF, E., 'The Finn in America,' *Geogr. Rev.*, **6**, 1918, 185–214.

5. CHISHOLM, G. G., 'Generalizations in Geography, especially in Human Geography,' *Scot. Geogr. Mag.*, **32**, 1916, 507–19.

6. JAMES, P. E. and JONES, C. F. (ed.), *American Geography, Inventory and Prospect*, Syracuse 1954: CHABOT, G. and CLOZIER, R., *La Géographie Française au milieu du XXe siècle*, Paris, 1957.

7. Noted in the Obituaries in the *Geogr. Rev.*, **41**, 1951, 51.

8. See ref. 3, page 3.

9. JOERG, W. L. G., 'Recent Geographical Work in Europe,' *Geogr. Rev.*, **12**, 1922, 431–84.

10. WALLIS, J. P. R. (ed.). *The Zambezi Expedition of David Livingstone* 1858–63, London, 1956, vol. 1, 88.

11. Quoted in the obituary by E. W. GILBERT, *Geogr. J.*, **110**, 1947, 94–9.

12. *Geogr. Teach.*, **9**, 1917–18, 195–6.

13. KELTIE, J. S., *Geogr. Teach.*, **7**, 1913–14, 215–25.

14. *Geogr. J.*, **2**, 1893, 162–4, note on 'Geography at the Forty-Second German Educational Congress'.

15. *Supp. Papers*, R. G. S., **1**, 1886, KELTIE, J. SCOTT, 'Geographical Education—Report to the Council of the Royal Geographical Society,' 459, 540. The whole report covers pp. 443–594.

16. *Scot. Geogr. Mag.*, **32**, 1916, 490–1.

17. THORNTHWAITE, C. W., 'Problems in the Classification of Climates,' *Geogr. Rev.*, **33**, 1943, 233–55, and other papers, including those in ibid. **21**, 1931, 633–55; **23**, 1933, 433–40; **38**, 1948, 55–94.

18. KELTIE, J. SCOTT, 'Thirty Years Progress in Geographical Education,' *Geogr. Teach.*, **7**, 1913–14, 224.

19. E. DE MARTONNE'S *Europe Centrale* in *Géogr. Univ.* 2 vols., Paris, 1930 and 1931, is generally regarded as one of the finest contributions to this great French series.

20. There were four editions of I. BOWMAN'S *New World*, 1921, 1923, 1924, 1928, but by the 1930's some students found its prophecies irritating as the world international and especially the economic situation had greatly changed. Much later, however, many read it again as an interesting product of its time.

21. GOTTMANN, J., 'Geography and the United Nations,' *Scot. Geogr. Mag.*, **66**, 1950, 134.

22. Ministry of Housing and Local Government, *Report*, 1959, 199.

CHAPTER TWO

1. *J. R. Geogr. Soc.*, **28**, 1858, cxxiii–ccxviii.

2. ibid. cxci.

3. *Supp. Papers*, R. G. S., **3**, 1893, 162.

4. *J. R. Geogr. Soc.*, **30**, 1860, cxxviii.

5. ibid. clv.
6. ibid. clxxvi–clxxix.
7. ibid. clxxvi.
8. ibid. cxv–cxci.
9. CRONE, G. R. and SKELTON, R. A., 'English Collections of Voyages and Travels, 1625–1846', in *Richard Hakluyt and His Successors* (Hakluyt Society, Second Series **93,** 1946), 78.
10. TAYLOR, E. G. R., *Tudor Geography*, 1485–1583, London, 1930: *Late Tudor and Early Stuart Geography*, 1583–1650, London, 1934.
11. See refs. to chapter 1, no. 1.
12. BAKER, J. N. L., 'Geography and Its History.' *Advanc. Sci.* no. **46** 1955, 198.
13. Originally given as President of Section E, at the British Association, 1901: British Association, *Report*, 1901, 701 (address 698–714).
14. This work was originally published in 1894 and has gone through twenty-five editions.
15. J. L. MYRES, *The Dawn of History*, London, 1920: *Geographical History in Greek Lands*, Oxford, 1953: Zimmern, A. E., *The Greek Commonwealth*, Oxford, 1911.
16. MARKHAM, C. R. (Sir Clements Markham), op. cit. in note 3.
17. *D.N.B.*, HOOKER, 2nd Supp., vol. 2, 294–9.
18. Obit. in MURCHISON, R. I., *J. R. Geogr. Soc.*, **28,** 1858, cxxiii–cxxvi: the population estimates are discussed in Connell, K. H., *The Population of Ireland*, 1750–1845, Oxford, 1950, 19, 256–7.
19. Obit. in *J. R. Geogr. Soc.*, **35,** 1865, cxxviii–cxxxi: and *D.N.B.*
20. ibid. cxxx.
21. CRONE, G. R., *Maps and Their Makers*, London, 1953, 155–6.
22. FREEMAN, T. W., *Pre-Famine Ireland*, Manchester, 1957, 317.
23. Obit. in *J. R. Geogr. Soc.*, **23,** 1853, lxviii–lxx, and the biography Portlock, J. E., *Memoir of the Life of Major-General Colby*, London, 1869.
24. Obit. in *D.N.B.*, vol. 16, 41–5.
25. Obit. in *J. R. Geogr. Soc.*, **34,** 1864, cxv–cxvii.
26. *D.N.B.* vol. 32, 143–5: *J. R. Geogr. Soc.*, **50,** 1880, clxviii. See also the *Report of the Commissioners Appointed to Inquire into the Facts Relating to the Ordnance Memoir of Ireland*, London, 1844.
27. *Census of Ireland*, 1841, Dublin, 1843. On these maps, see note in *Geogr. J.*, **122,** 1956, 129–31.
28. *Second Report of the Commissioners Appointed to Consider and Recommend a General System of Railways for Ireland*. Parliamentary Papers, vol. **35,** 449, 1837–8. Interesting evidence on this work was given by T. A. LARCOM in the *Report* mentioned in note 26, pp. 569–70.

29. ROBINSON, A. H., 'The Maps of Henry Drury Harness,' *Geogr. J.*, **121**, 1955, 440–50.
30. This was made plain at the inquiry noted in the *Report*, Ref. 26: for example, on p. iv.
31. LEWIS, S., *A Topographical Dictionary of England*, 4 vols., London, 1831; *A Topographical Dictionary of Ireland*, 3 vols., 1837. The *D.N.B.* account of his work is not entirely accurate, see ref. 22, 315.
32. SINCLAIR, J., *The Statistical Account of Scotland*, Edinburgh, 1791–9: also *The New Statistical Account of Scotland*, 15 vols., Edinburgh 1845.
33. The work of Ritter is thoroughly discussed, with bibliography, in Hartshorne, R., *The Nature of Geography*, 1939, 48–84.
34. *Europa, ein Geographisch-Historisch-Statisches Gemälde*, 2 vols., Frankfurt-a-Main, 1804, 1807.
35. *Scot. Geogr. Mag.*, **75**, 1959, 156.
36. BAKER, J. N. L., op. cit. in ref. 12, 189.
37. Published in 5 volumes as *Kosmos: Entwurf einer Physischen Weltbeschreiburg*, Stuttgart 1845–62: a part was translated into English, by C. E. OTTÉ and W. S. DALLAS, appeared as *Cosmos: a Sketch of a Physical Description of the Universe*, 5 vols., London, 1849–58. The work was translated into several languages: see Bruhns, K., *Alexander von Humboldt* (3 vols.), Leipzig 1872, esp. vol. II, 523–5.
38. CORTAMBERT, P. F. Eugène, *Physiographie*, 1836.
39. HUXLEY, T. H., *Physiography: An Introduction to the Study of Nature*, London, 1877.
40. BAKER, J. N. L., 'Mary Somerville and Geography in England,' *Geogr. J.*, **111**, 1948, 207–22. The preface to Mrs Somerville's book *On the Connexion of the Physical Sciences*, is dated 1834 and the third edition 'again carefully revised', was published by John Murray in 1836. The first edition of *Physical Geography* in two volumes, was published in London by John Murray in 1848, there were later editions in 1849, 1851, 1858, 1862, 1870 and 1877.
41. HARTSHORNE, op. cit. in ref. 33, 77.
42. ibid. 67.
43. ibid. 64.
44. ibid. 68.
45. ibid. 79.
46. LEIGHLY, J., 'Methodological Controversies in Nineteenth-Century German Geography,' *Ann. Ass. Amer. Geogr.*, **28**, 1938, 241.
47. HUMBOLDT, A. VON, *Cosmos*, trans. E. C. OTTÉ, London, 1849, vol. 1, 22, and ix–xiv.

48. VIDAL DE LA BLACHE, *Principles of Human Geography*, London, 1926, translated by M. T. BINGHAM, 6–7. The date of Berghaus is wrongly given as 1836.

49. HUMBOLDT, op. cit., vol. I, 22.

50. ibid. 9–10.

51. See translator's preface in Humboldt, op. cit., note 47.

52. KRAMER, F. L., 'A Note on Carl Ritter,' *Geogr. Rev.*, **49,** 1959, 406–09.

53. GUYOT, A. H., 'Carl Ritter', *J. Amer. Geogr. and Stat. Soc.*, **2** (part 1) 1860, 25–63, esp. pp. 40–1.

54. SINNHUBER, K. A., 'Carl Ritter,' *Scot. Geogr. Mag.*, **75,** 1959, 159.

55. BÖGEKAMP, H., 'An Account of Professor Ritter's Geographical Labours,' in Gage, W. L. (trans.), *Geographical Studies by the late Professor Carl Ritter of Berlin*, New York, 1861, 33–51.

56. GUYOT, A. H., *The Earth and Man* first appeared in America in 1863. Unfortunately it was a dull work: on Guyot, see pp. 40–1.

57. NEWBIGIN, M. I., *The Mediterranean Lands*, London, 1924, esp. p. 21.

58. First published Oxford, 1902; 2nd edition 1907, with several reprints.

59. GALLOIS, L., *Régions Naturelles et Noms de Pays*, Paris, 1908.

60. ibid. 7.

61. ibid. 16, 19–20.

62. ibid. 27.

63. This famous work was first published in 1903 as part of Lavisse's *Histoire de France*.

64. BAKER, J. N. L., 'Mary Somerville and Geography in England,' *Geogr. J.*, **III,** 1948, 207–08.

65. ibid. 209. See ref. 40.

66. ibid. 216.

67. ibid. 213.

68. KELTIE, J. SCOTT, Report to the Council of the Royal Geographical Society, *Supp. Papers, R. G. S.*, **I,** 1886, 450–1.

69. ibid. 466.

70. ibid. 546–7.

71. ibid. 547–8.

72. (a) HUGHES, W., *Geography in its Relation to History:* a lecture delivered at the Birkbeck Institution, 1870: (b) *Geography in Relation to Physical Science*, Bedford College, London, 1870.

73. ibid. (b), 1, 5–6.

74. ibid. 12.

75. ibid. 17.

76. ibid. 1.

77. See note 72(a) 9. Hughes was an admirer of Harriet Martineau's work.

78. HUGHES, W., *An Atlas of Classical Geography*, ed. G. Long, London, 1854. It has twenty-four plates.

79. FAURE, C., 'Notice sur Arnold Guyot,' *Globe*, **23**, Geneva, 1884.

80. ibid. 30.

81. GUYOT, A., *The Earth of Man, or Physical Geography in its Relation to the History of Mankind*, London, 1852, 6th edn. 1863. The lectures were given in French and translated by C. C. Felton. Other works of Guyot included an article 'on the Appalachian mountain system', *American Journal of Science*, New Haven, Connecticut, **31**, 1861, 157–87, and *A Collection of Meteorological Tables, with Other Tables Useful in Practical Meteorology*, Washington, 1852, and later editions.

82. LOWENTHAL, D., *George Perkins Marsh, Versatile Vermonter*, New York, 1958.

83. ibid. 248.

84. ibid. 246.

85. ibid. 214–15.

86. MARSH, G. P., *The Earth and Man*, New York, 1864.

87. LOWENTHAL, op. cit. 270.

88. ibid. 271.

89. ibid. 268.

90. ibid. 305–09.

91. MARKHAM, C. R., 'The Fifty Year's Work of the Royal Geographical Society,' *J. R. Geogr. Soc.*, **50**, 1879–80, 125–6.

92. ibid. 118.

93. See references to chapter 10.

94. Obit. in *J. R. Geogr. Soc.*, **43**, 1873, clx.–clxiii.

95. Obit. in *J. R. Geogr. Soc.*, **44**, 1874, cxxxi–cxxxiii.

96. Obit. in *Proc. R. Geogr. Soc.*, **I**, 1879, 133–4.

97. FREEMAN, T. W., *Ireland*, 1950, 135.

98. GILBERT, E. W., 'Pioneer Maps of Health and Disease in England,' *Geogr. J.*, **124**, 1958, 172–83.

99. See note 48; note 95; and JOHNSTON, T. B., *Notes on the Geographical Labours of A. Keith Johnston*, n.d. (in Royal Geographical Society); obit. *J. R. Geogr. Soc.*, **42**, 1872, clxi–clxiii. The globe mentioned in this paragraph may be seen at the Royal Geographical Society, London.

100. GILBERT, op. cit., 178.

101. PORTLOCK, J. E., *Memoir of the Life of Major-General Colby*, London, 1869, esp. 303.

102. *Census of Ireland*, 1881, 77.

103. *Report of the Commissioners . . . Ordnance Memoir of Ireland*, London, 1844 (see note 26). Sir Robert Peel's comment is on p. iv and Sir Robert Kane's comments on pp. 66 and 78.

104. McKay, D. V., 'Colonialism in the French Geographical Movement, 1871–81,' *Geogr. Rev.*, **33**, 1943, 214–32.
105. Faure, C., 'Le Progrès de l'enseignement de la Géographie en France,' *Bulletin de la Société Neuchâteloise de Géographie*, **6**, 1891, 96–125, esp. 101–15. This spirited article ends with the word *Excelsior*.
106. Obit. of St. Martin, V. de, *Comptes Rendus des Séances de al Socéité de Géographie*, Paris, 1897, 5–7.
107. Obit., *Ann. Géogr.*, **14**, 1905, 373–4.
108. *Bulletin de la Société de Géographie*, Paris, 6th series, tome 11, 1876, 465–534.
109. ibid. Derrécagaix, P. C., 'Notice sur les Basques,' 401–38. The comment quoted is on p. 401.
110. *Bulletin de la Société de Géographie de Marseille*, **1**, 1877, 6 (in preface).

CHAPTER THREE

1. *Geogr. J.*, **7**, 1896, 310–11.
2. *J. R. Geogr. Soc.*, **50**, 1880, 7–10.
3. Howarth, O. J. R., The Centenary of Section E (Geography), *Advanc. Sci.*, no. **30**, 1951, 151, 152.
4. Wright, J. K., 'The Field of the Geographical Society,' in *Geography in the Twentieth Century*, ed. G. Taylor, London, 1951, 548.
5. Mill, H. R., *The Record of the Royal Geographical Society*, 1830–1930, London, 1930, 44, 146.
6. McKay, D. V., 'Colonialism in the French Geographical Movement, 1871–81,' *Geogr. Rev.*, **33**, 1943, 214–17.
7. ibid. 226.
8. ibid. 220.
9. ibid. 222–3.
10. ibid. 227–8.
11. ibid. 229.
12. Many of these societies have been traced through their publications, in the library of the Royal Geographical Society, London. Some are mentioned in contemporary journals. There is also valuable information in Wright, J. K. and Platt, Elizabeth, *Aids to Geographical Research*, New York, 1923, 1947, particularly in the 1947 edition, section 'Regional Aids and General Geographical Periodicals.'
13. 'Geographisches Jahrbuch,' 1885, quoted in *Scot. Geogr. Mag.*, **2**, 1886, 182.
14. Quoted in *Geogr. J.*, **9**, 1897, 451.

15. WRIGHT, J. K., *Geography in the Making—the American Geographical Society*, 1851–1951, New York, 1952, esp. 25–30, 70.
16. KANE, E. K., 'Access to an Open Polar Sea along a North American Meridian,' *Bull. Amer. geogr. stat. Soc.*, **1**, 1852, 85–104; *Smithsonian Contributions to Knowledge*, New York, Astronomical Observations in the Arctic Seas, May 1860 (49 pp.), Tidal Observations in the Arctic Seas, October 1860 (83 pp.); Arctic Explorations in the Years 1853, '54, '55, 2 vols. Philadelphia 1856–7. A contemporary work is Elder, W., *Biography of Elisha Kent Kane*, Philadelphia, 1858, and a modern life in Mirsky, J., *Elisha Kent Kane and the Seafaring Frontier*, Boston, 1954.
17. WRIGHT, J. K., op cit. 27–9: HOPKINS, E. A., 'Memoir on the Geography, History and Productions, and Trade of Paraguay.' *Bull. Amer. geogr. stat. Soc.*, **1**, 1852, 14–46: MURCHISON, R. I., 'Address', *J. R. Geogr. Soc.*, **23**, 1853, cxviii.
18. WRIGHT, J. K., op. cit. 86.
19. ibid. 86.
20. ibid. 87–8.
21. MILL, H. R., op. cit. (ref. 5) 22–34.
22. ibid. 44.
23. ibid. 244–6.
24. BAKER, J. N. L., 'Mary Somerville and Geography in England', *Geogr. J.*, **III**, 1948, 207–22.
25. GAGE, W. L. (trans.), *Geographical Studies by the late Prof. Ritter of Berlin*, New York, 1861, iv.
26. *J. R. Geogr. Soc.*, **43**, 1873, clxi–clxiii.
27. *Proc. R. Geogr. Soc.*, **5**, 1883, 387.
28. ibid. **6**, 1884, 373.
29. ibid. **7**, 1885, 428.
30. ibid. **7**, 1885, 418.
31. ibid. **7**, 1885, 424–5, 607.
32. ibid. **7**, 1885, 746–7. Early numbers of the Victoria and New South Wales branches exist in the Royal Geographical Society, London, as well as a work called *Special record of the proceedings of the Geographical Society of Australasia, in fitting out and starting the exploratory expedition to New Guinea*, compiled by E. Pulsford, Sydney 1885. The first number of the Adelaide journal appears as *Proceedings of the Geographical Society of Australia, South Australia branch*, 1886, and vol. 1 of the Queensland branch is dated 1886 also. In 1928 the Geographical Society of New South Wales launched the *Australian Geographer* with an opening article by Griffith Taylor.
33. ibid. **7**, 1885, 244. The *Transactions of the Geographical Society of*

Quebec, with a French title also, first appeared in 1880: there were papers in both languages.

34. ibid. **7,** 1885, 745.

35. ibid. **7,** 1885, 428.

36. *Scot. Geogr. Mag.*, **1,** 1885, 1–17.

37. *J. Manch. Geogr. Soc.*, **1,** 1885, 1–25. If the date of foundation of a society dates from the inaugural meeting, Manchester had its society after Scotland, but if from the visit of H. M. Stanley, the Scottish is the junior society by six weeks.

38. MILL, H. R., *An Autobiography*, London, 1951, 82–8: also MILL, H. R. in *Geogr. J.*, **5,** 1895, 360–8.

39. TAIT, J. B., 'Oceanography: Scotland's Interest in its Progress,' *Scot. Geogr. Mag.*, **61,** 1945, 1–8, *J. R. Geogr. Soc.*, **44,** 1874, clvi–clxiii; **45,** 1875, clxi–clxiii; **46,** 1876, clxi–clxvii; **47,** 1877, clxii–clxiv; MILL, H. R., *An Autobiography*, London, 1951, 42–4.

40. MANLEY, G., *Climate and the British Scene*, London, 1952, 188–9, 207, 210–11.

41. Obits. *J. R. Geogr. Soc.*, **42,** 1872, clxi–clxiii, *Proc. R. Geogr. Soc.*, **1,** 1879, 598–600; JOHNSTON, T. B., *Notes on the Geographical Labours of A. Keith Johnston*, n.d. but *c.* 1872.

42. *Scot. Geogr. Mag.*, **1,** 1885, 119–23.

43. ibid. 124.

44. Available in the library of the Royal Geographical Society.

45. *Ann. Géogr.*, **1,** i–iii.

46. *Geogr. J.*, **1,** 1893, 264–5.

47. *Scot. Geogr. Mag.*, **1,** 1885, 487–96.

48. *Geogr. J.*, **2,** 1893, 165–6: **4,** 1894, 237–46; **22,** 1903, 237–69, 521–41; **23,** 1904, 32–61.

49. WEBSTER, op. cit. in ref. 48. The articles on 'onomatology' include one by J. J. Egli (1825–96), 'the leading European authority on the subject', *Scot. Geogr. Mag.*, **1,** 1885, 422–8, and others in vol. 2.

50. WEBSTER, ibid. 495.

51. MILL, H. R., *Geogr. J.*, **7,** 1896, 345–65; **15,** 1900, 205–27, 353–78; MILL, H. R., op. cit. in ref. 39, 98.

52. MILL, H. R., *Geogr. Teach.*, **11,** 1921–2, 7–19.

53. *Scot. Geogr. Mag.*, **2,** 1886, 733–9.

54. ibid. **3,** 1887, 161–9.

55. *Geogr. J.*, **2,** 1893, 481–2.

56. *Proc. R. Geogr. Soc.*, **9,** 1887, 141.

57. *Geogr. J.*, **6,** 1895, 367.

58. ibid. 368–76.

59. *National Geographic Magazine*, **1,** New York, 1889.

60. WRIGHT, J. K., op. cit. in ref. 15, 119.

61. Available at the Royal Geographical Society.
62. *Geogr. J.*, **5**, 1895, 369–73; the resolutions of the Congress are noted in ibid. **6**, 1895, 269–74.
63. PENCK, A., 'The Construction of a Map of the World on a Scale of 1:1,000,000,' *Geogr. J.*, **1**, 1893, 253–61. See also *Geogr. J.*, **5**, 1896, 275–6; **6**, 1896, 272; **32**, 1908, 373–5 (report by G. G. Chisholm on further progress at the 9th Congress in Geneva); *Geogr. Rev.*, **41**, 1951, 25–7 on the work of I. Bowman.

CHAPTER FOUR

1. *Geogr. J.*, **2**, 1893, 358.
2. ibid. **26**, 1905, 593.
3. ibid. **10**, 1897, 14–15.
4. FLEURE, H. J., 'Sixty Years of Geography and Education,' *Geography*, **38**, 1953, esp. 236–45.
5. *Geogr. J.*, **7**, 1896, 208.
6. HERBERTSON, A. J., 'Recent Discussions on the Scope and Educational Approaches of Geography,' *Geogr. J.*, **24**, 1904, 417–27.
7. The famous article by A. J. HERBERTSON, 'The Major Natural Regions,' appeared in *Geogr. J.*, **25**, 1905, 300–10: on mammals, there was a long series of regional articles by W. L. SCLATER, in *Geogr. J.*, **5** onwards. Perusal of the journals of the years before 1914 will show that regional classification was a constant concern.
8. TAYLOR, G., 'The Physiographic Control of Australian Environment,' *Geogr. J.*, **53**, 1939, 172–92, and articles in *Scot. Geogr. Mag.*, **48**, 1932, 1–20, 65–78. Taylor's writings are numerous, and include *Australian Meteorology*, Oxford, 1920 and *Australia*, London, 1940, with later editions in 1943, 1951 and reprintings. An early work is 'The Australian Environment (especially as controlled by rainfall)', *Commonwealth of Australia, Advisory Council of Science and Industry, Memoir I*, Melbourne, 1918.
9. *Geogr. J.*, **2**, 1893, 355.
10. DAVIS, W. M., 'Geography in the United States,' *Proceedings of the American Association for the Advancement of Science*, **53**, 1904, 1–32, esp. 28–30. Davis wrote a large number of papers on education.
11. *Geogr. J.*, **24**, 1904, 418.
12. BAULIG, H., 'William Morris Davis: Master of Method,' *Ann. Ass. Amer. Geogr.*, **40**, 1950, 188. The story that Davis might lose his job through lack of published material is told by L. MARTIN in op. cit. 178.
13. DAVIS, W. M., *Ann. Ass. Amer. Geogr.*, **14**, 1924, 171–2.

14. PEATTIE R., quoted in *Ann. Ass. Amer. Geogr.*, **40,** 1950, 178.
15. LEIGHLY, J., *Ann. Ass. Amer. Geogr.*, **45,** 1955, 309.
16. See ref. 7.
17. Summary in *Geogr. J.*, **1909,** 329–30: see also original articles in *Petermanns Mitt.* Gotha, **54,** 1908, 147–60, 182–8, and 'Physiographie und Vergleichende Landschaftsgeographie' in the *Report of the 1913 Geographical Congress* at Rome (*Atti del X Congresso Internazionale di Geographia, Roma*), 1915, 755–86.
18. MILL, H. R., 'The Value of Regional Geography,' *Geogr. Teach.* **11,** 1921–2, 63–75.
19. KENDREW, W. G., *The Climates of the Continents*, Oxford, 1937, 271.
20. RITTER, C., *Comparative Geography*, trans. W. L. Gage, New York, 1865, xv.
21. DAVIS, W. M., 'The Physical Geography of the Lands,' *Popular Science Monthly*, **58,** 1900, 157–70, esp. 158: DALY, R. A., *Biogr. Mem. Nat. Acad. Sci.*, **23,** 1945, 270–1.
22. DAVIS, W. M., 'The Progress of Geography in the Schools,' originally published in the *First Year Book, National Society for the Scientific Study of Education*, 1902, part 2, 7–49, and reprinted in *Geographical Essays*, ed. D. W. Johnson, 1909, 1954. See esp. pp. 32–8.
23. RITTER, C., op. cit. in ref. 20, xvi–xvii.
24. WOOD, A., *The Groundnut Affair*, London, 1950.
25. BOWMAN, I., *The Pioneer Fringe*, American Geographical Society pub. **13,** 1931, 42–3. This work is somewhat optimistic. See also Bowman's *Limits of Land Settlement*, New York, 1937.
26. BRIGHAM, A. P., 'Problems of Geographic Influence,' *Ann. Ass. Amer. Geogr.*, **5,** 1915, 23.
27. BARNES, H. E., *The New History and the Social Sciences*, New York, 1925, 49, says that H. T. Buckle was much misrepresented. He tried to make history an exact science of culture development and attempted to estimate the importance of external or geographical factors on civilization: for example he considered the effect of climate, soil and food on human cultures. He thought that great scenery could arrest civilization by mentally paralysing people, as in north India, but a variegated scenery, such as that of the Greek peninsula, was well adapted to produce a progressive civilization.
28. OGDEN, H. W., 'The Geographical Basis of the Lancashire Cotton Industry,' *J. Manch. Geogr. Soc.*, **43,** 1927, 8–28. This paper argues that the localization of the cotton industry was due to the existence of plentiful supplies of soft water from Pennine streams, and not as previously supposed to a humid climate which reduced the breakage of fibres.

29. Obits., *Geogr. Rev.*, **22,** 1932, 500–01: *Ann. Ass. Amer. Geogr.*, **23,** 1933, 229–40 (by C. C. Colby).

30. SEMPLE, E. C., *Influences of Geographic Environment*, New York, 1911, v–viii, 51–2. The idea of a geographical core for a state is favoured by some workers now, cf. pp. 210, 215.

31. ibid. vii.

32. ibid. 33–4, 36.

33. ibid. 35.

34. ibid. 37–8.

35. ibid. 41.

36. ibid. 43.

37. ibid. 42.

38. HUNTINGTON, E., 'Climatic Variations and Economic Cycles,' *Geogr. Rev.*, **1,** 1916, 197–8.

39. ibid. 195.

40. An excellent summary is given in S. S. VISHER, 'Memoir to Ellsworth Huntington, 1876–1947,' *Ann. Ass. Amer. Geogr.*, **38,** 1948, 38–50.

41. MILL, H. R., 'A Fragment of the Geography of England: South-West Sussex,' *Geogr. J.*, **15,** 1900, 205–27, 353–78. *An Autobiography*, 1951, 98.

42. MILL, H. R., 'England and Wales Viewed Geographically,' *Geogr. J.*, **24,** 1904, 621–36.

43. HERBERTSON, A. J., 'The Major Natural Regions of the World,' *Geogr. J.*, **25,** 1905, 300–10.

44. HERBERTSON, A. J., *Geogr. Teach.*, **7,** 1913–14, 158–63. It is generally believed that Herbertson intended to devote his later years to the study of human geography.

45. *Geogr. J.*, **23,** 1904, 111–14.

46. ibid. 114.

47. UNSTEAD, J. F. and TAYLOR, E. G. R., *General and Regional Geography*, London, 1910, esp. pp. 237, 331 (10th ed. 1927): DRYER, C. R., *High School Geography*, 1911.

48. UNSTEAD, J. F., 'A System of Regional Geography,' *Geography*, **18,** 1933, 175–87. See also A. J. Herbertson's comments on 'Types and orders of natural regions', *Geogr. Teach.*, **7,** 1913–14, 160–3.

49. POWELL, J. W., 'Physiographic Regions of the United States,' *National Geographic Society Monographs*, **1,** 1895, 65–100 (coloured map pp. 98–9). This monograph volume contains several important papers on the geomorphology of North America by W. M. DAVIS, G. K. GILBERT, I. C. RUSSELL, N. S. SHALER, BAILEY WILLIS and others.

50. FENNEMAN, N. M., obit. *Geogr. Rev.*, **35,** 1945, 682; *Ann. Ass.*

Amer. Geogr., **35**, 1945, 181–9. His papers on the physiographic regions of the United States appeared in *Ann. Ass. Amer. Geogr.*, **4**, 1914, 84–134 and **6**, 1916, 19–98. These were preliminary to his famous books on the physiography of the United States, published 1931 and 1938.

51. JOERG, W. L. G., 'The Subdivision of North America into Natural Regions: A Preliminary Inquiry,' *Ann. Ass. Amer. Geogr.*, **4**, 1914, 55–83.

52. SMITH, W. G. and RANKIN, W. M. in *Geogr. J.*, **22**, 1903, 149–94: PETHYBRIDGE, G. H. and PRAEGER, R. L., 'The Vegetation of the District Lying South of Dublin,' *Proceedings of the Royal Irish Academy*, **25**, 1905, 124–80. There were other similar articles written at this time—for example by F. J. LEWIS in *Geogr. J.*, **23**, 1904, on the northern Pennines.

53. Published at Oxford: review in *Geogr. J.*, **24**, 1904, 217.

54. *Scot. Geogr. Mag.*, **24**, 1908, 370.

55. *Geogr. J.*, **27**, 1906, 553–6.

56. A critical assessment is given by Lewis Mumford, 'Patrick Geddes, Victor Branford, and Applied Sociology in England,' in Barnes, H. E. (ed.), *An Introduction to the History of Sociology*, Chicago, 1948, 677–95. The suggestion of a museum is given in *Scot. Geogr. Mag.*, **18**, 1902, 1902, 142–4 and design, opp. 168. See also Boardman, P., *Patrick Geddes: Maker of the Future*, Chapel Hill, 1944 (though interesting and informative, this work suffers in places from a somewhat melodramatic style); MAIRET, P., *Pioneer of Sociology*, London, 1957.

57. FLEURE, H. J., 'Human Regions,' *Scot. Geogr. Mag.*, **35**, 1919, 94–105.

58. The great contribution of the French to regional geography is noted in HARRISON CHURCH, J., 'The French School of Geography,' in TAYLOR, G., *Geography in the Twentieth Century*, London, 1950, esp. 73–9.

59. Quoted in HARTSHORNE, R., 'The Nature of Geography,' *Ann. Ass. Amer. Geogr.*, **29**, 1939, 140.

60. DRYER, C. R., 'Natural Economic Regions,' *Ann. Ass. Amer. Geogr.*, **5**, 1915, 121–5.

61. This work is still used: from 1928 it was revised by L. D. Stamp and from 1954 by L. D. STAMP and S. C. GILMOUR. The 1954 issue, the fourteenth edition, is described as 'entirely re-written'.

62. CHISHOLM, G. G., 'Economic Geography,' *Scot. Geogr. Mag.*, **24**, 1908, 113–32.

63. CHISHOLM, G. G., 'The Meaning and Scope of Geography,' *Scot. Geogr. Mag.*, **24**, 1908, 561–75.

64. ibid. ref. 62, 117–18.

65. CHISHOLM, G. G., *Handbook of Commercial Geography*, 4th edn., London, 1903.
66. VIDAL DE LA BLACHE, *États et Nations de l'Europe, Autour de la France*, 1891, v.
67. MACKINDER, H. J., 'The Geographical Pivot of History,' *Geogr. J.*, **23**, 1904, 421–37.
68. HARTSHORNE, R. in *American Geography: Inventory and Prospect*, ed. JAMES P. E. and JONES C. F., Syracuse, 1954, 174.
69. Published in London and New York, 1919 and 1942: in New York (exactly as in 1919) in 1942.
70. MACKINDER, 1919 op. cit. 64–5.
71. ibid. 74–83.
72. MACKINDER's revised views appeared in *Foreign Affairs*, New York, **21**, 1942–3, 595–605. They have been extensively discussed: see, for example, JONES S. B., *Geogr. Rev.*, **45**, 1955, 317–18, 494–508.
73. CHISHOLM, G. G., *A Smaller Commercial Geography*, London, 1890, 1905.
74. The *Oxford Survey of the British Empire*, ed. A. J. HERBERTSON and O. J. R. HOWARTH, Oxford, 1914, is a compilative work.
75. McFARLANE, J., *Economic Geography*, 4th edn., London, 1937, vii. This work has been reissued with revisions by C. F. R. GULLICK.

CHAPTER FIVE

1. MANLEY, G., *Climate and the British Scene*, London, 1952.
2. TANSLEY, A. G., *The British Islands and Their Vegetation*, Cambridge, 1939.
3. On Cvijić, see *Note sur les Travaux Scientifiques de M. Jovan Cvijić* (in Royal Geographical Society), n.d.
4. Obits., *Ann. Géogr.*, **65**, 1956, 1–14 (by A. Cholley); *Geogr. J.*, **121**, 1955, 547–9; *Geogr. Rev.*, **46**, 1956, 277–9.
5. This print is well made by G. M. WRIGLEY, *Geogr. Rev.*, **41**, 1951, 7–16.
6. SÖLCH, J., *Die Britischen Inseln*, I, *England und Wales*, 1951, II, *Schottland und Irland*, 1952.
7. *Ann. Ass. Amer. Geogr.*, **40**, 1950, 172–236, by various authors.
8. DAVIS, W. M., 'The Explanatory Description of Landforms,' in *Recueil de travaux offert à Jovan Cvijić par ses amis et collaborateurs a l'occasion de ses trente-cinq ans de travail scientifique*, Belgrade, 1924, 1–50.
9. DAVIS, W. M., 'The Progress of Geography in the United States,' *Ann. Ass. Amer. Geogr.*, **14**, 1924, 159–215, esp. 201.

10. DAVIS, W. M., 'The Long Beach Earthquake,' *Geogr. Rev.*, **24,** 1934, 1–11.
11. LOBECK, A. K., *Geomorphology: An Introduction to the Study of Landscapes*, New York and London, 1939.
12. COTTON, C. A., *Geomorphology of New Zealand, Part I Systematic*, Wellington, N.Z., 1922.
13. BRYAN, K., 'Place of Geomorphology in the Geographic Sciences,' *Ann. Ass. Amer. Geogr.*, **40,** 1950, 197.
14. PLATT, R. B., *Finland and Its Geography*, New York, 1955.
15. For a general assessment, see DALY, R. A., 'Biographical Memoir of William Morris Davis, 1850–1934,' *Biogr. Mem. Nat. Acad. Sci.*, 23, 1945, 263-303. DALY notes that DAVIS published his *Elementary Meteorology* in 1894, having written forty papers on meteorological topics from 1884–93: this work was passed on to R. de Courcy Ward. Daly gives a full bibliography and notes (p. 277) that the only comprehensive statement of his matured philosophy was published in German, *Die Erklarende Beschreibung der Landformen*, 1912.
16. BRYAN, L., *Ann. Ass. Amer. Geogr.*, **40,** 1950, 202.
17. STRAHLER, A. N., ibid. 209–13.
18. As, for example, in 'The Geographical Cycle', originally published in *Geogr. J.*, **14,** 1899, 481–504, and reprinted in *Geographical Essays*, ed. D. W. JOHNSON, 1909, 1954 (Dover edition).
19. D. W. JOHNSON, op. cit. 1954, 690–724.
20. JOHNSON, D. W., *Shore Processes and Shoreline Development*, New York, 1919.
21. *Geographical Essays*, op. cit. esp. nos. 19, 20, 22.
22. JOHNSON, D. W., *Stream Structure on the Atlantic Slope*, New York, 1931.
23. DAVIS, W. M. in *Ann. Ass. Amer. Geogr.*, **14,** 1924, 188, 191–2.
24. ibid. 182–4.
25. ibid. 192.
26. ibid. 260.
27. JUKES, J. B., 'On the Formation of some River Valleys in the · South of Ireland,' *Quarterly Journal of the Geological Society*, **18,** London, 1862, 378–403. Though accepted by most later workers, Jukes's theories are now questioned, due to an unexpected geological find in Co. Kerry.
28. DAVIS, op. cit. in ref. 23, 182–3.
29. BRYAN, K., op. cit. in ref. 16, 203–06.
30. WOOLDRIDGE, S. W. and MORGAN, R. S., *The Physical Basis of Geography*, London, 1937, 175, 190.
31. BAULIG, H., *Ann. Ass. Amer. Geogr.*, **40,** 1950, 190–3; also in *J. Geomorph.*, **1,** 1938, 228.
32. *J. Geomorph.*, **1,** 1938, 40.

33. In *J. Geomorph.*, **3**, 1940, 180, there is an abstract of two papers by A. A. MILLER, in *Proceedings of the Royal Irish Academy*, **45**, 1939, 321–54 (on the same area as that treated in the Jukes 1862 paper, ref. 27) and *Proceedings of the Yorkshire Geological Society*, **24**, Leeds, 1938, 31–59.

34. RICH, J. L. 'Recognition and Significance of Multiple Erosion Surfaces,' *Bulletin of the Geological Society of America*, **49**, Rochester, N.Y., 1938, 1695–1722; summary in *J. Geomorph.*, **2**, 1939, 265–7.

35. 'Glaciers and Climate, Geophysical and Geomorphological Essays presented to H. W.: son Ahlmann,' *Geografiska Annaler*, Stockholm, 1949, 21–24, 54–5, 212–21.

36. BAIRD, P. D. and LEWIS, W. V., 'The Cairngorm Floods,' 1956, *Scot. Geogr. Mag.*, **73**, 1957, 91–100.

37. op. cit. in ref. 34, 266.

38. The report by G. K. GILBERT on the Henry mountains appeared in 1879: see DAVIS, W. M. 'Biographical Memoir of Grove Karl Gilbert, 1843–1918,' *Biogr. Mem. Nat. Acad. Sci.*, **21**, 1927. W. M. DAVIS's essay on the cycle in an arid climate first appeared in *Journal of Geology*, **13**, Chicago, 1905, 150–63, and is reprinted in *Geographical Essays*, op. cit. 296–322.

39. DAVIS, W. M., *Geographical Essays*, 297. The work of Passarge is discussed by DAVIS in the essay.

40. *Richthofen, Letters from*, printed Shanghai 1870–2: there are descriptions of loess in the letters on Honan and Shansi, 1870, 9–10 and Chili, Shansi, Shensi, Sz'chwan (1872) 13–18. F. F. VON RICHTHOFEN's publications were numerous, and included, in 1900–02, 'Geomorphologische Studien aus Ostasien,' Berlin (*Königlich Preussischen Akademie der Wissenschaften zu Berlin*), 1900–03.

41. op. cit. in ref. 30, 286.

42. The most accessible English source in SANDERS, E. M., 'The Cycle of Erosion in a Karst Region (after Cvijic),' *Geogr. Rev.*, **11**, 1921, 593–604: the original appeared as 'Hydrographie souterraine et évolution morphologique du Karst,' *Recueil des Travaux de l'Institut de Géographie Alpine*, **6**, 1918, 375–426.

43. WILLETT, H. C. in op. cit. ref. 35, 296–9.

44. W. B. WRIGHT's *Quaternary Ice Age*, 2nd edn., 1936, is dedicated to R. L. Praeger as the 'discoverer of the climatic optimum'.

45. *Geogr. J.*, **125**, 1959, 230, 233–4, in article by M. MELLOR.

46. ibid. 234.

47. HALLAND-HANSEN, in op. cit. ref. 35, 75–80.

48. AHLMANN, H. W.: son, Glaciological Research on the North Atlantic Coasts, *Royal Geographical Society, Research Series*, **1**, 1948, 75–7.

49. op. cit. in ref. 35, 54–5.
50. WERENSKIOLD, W. in Ahlmann, op. cit. in ref. 47, 292–4.
51. AHLMANN, op. cit. in ref. 47, 75.
52. ibid. 74.
53. *Geogr. Rev.*, **40,** 1950, 179–223, articles on the Juneau icefield research project by W. O. Field and M. M. Miller, and on 'Glacier Fluctuation for Six Centuries in South-East Alaska ...' by D. B. LAWRENCE.
54. MANLEY, G. op. cit. in ref. 35, 179–93; and op. cit. in ref. 1,212.
55. FLINT, R. F., op. cit. in ref. 35, 56–74.
56. NIELSEN, N., *Atlas of Denmark*, Copenhagen, 1949.
57. COTTON, C. A., *Landscape as Developed by the Processes of Normal Erosion*, London, 1941; enlarged second ed., 1948; *Climatic Accidents in Landscape-making*, Christchurch and London, 1942, *Volcanoes as Landscape Forms*, Christchurch, 1944, 1952.
58. JOHNSON, D. W., *Shore Processes and Shoreline Development*, New York, 1919.
59. STEERS, J. A., *The Coastline of England and Wales*, Cambridge, 1946.
60. See ref. 56.
61. KING, C. A. M., *Beaches and Coasts*, London, 1959.
62. On the Exmoor flood, see *Geography*, **38,** 1953, 1–17: The east coast floods are discussed in *Geogr. J.*, **119,** 1953, 280–98 and, with those of the Netherlands also, in *Geography*, **38,** 1953, 132–89.

CHAPTER SIX

1. *Census*, Great Britain, 1851, vol. I, xix–xx, lxxxi.
2. GILBERT, E. W., 'Practical Regionalism in England and Wales,' *Geogr. J.*, **94,** 1939, 29–44; 'The Boundaries of Local Government Areas,' *Geogr. J.*, **111,** 1948, 172–206; 'The Idea of the Region,' *Geography*, **45,** 1960, 157–75.
3. FAWCETT, C. B., *Geogr. J.*, **111,** 1948, 201, noted that 'each important market town . . . became the centre of a Poor Law Union and the parishes linked with that market town were those whose farmer inhabitants came in to . . . that market town.'
4. *Geogr. Rev.*, **7,** 1919, 115–16.
5. *Géogr. Univ.*, 'France Economique et Humaine,' 1948, 847–8.
6. DICKINSON, R. E., 'The Economic Regions of Germany,' *Geogr. Rev.*, **28,** 1938, 609–26, and his later books on Germany.
7. Republished in 1960, London, with a new preface by W. G. East and S. W. Wooldridge, and some additions to the text.
8. op. cit. in ref. 1, xlv–xlix.
9. GEDDES, P., *Cities in Evolution*, London, 1915, expresses a liberal opinion of its time.

10. ODUM, H. W. and MOORE, H. E., *American Regionalism—a Cultural-Historical Approach to National Integration*, New York, 1938, 4.
11. ibid. 21.
12. GREEN, F. H. W., 'Urban Hinterlands in England and Wales: An Analysis of Bus Services,' *Geogr. J.*, **116**, 1950, 64–88; 'Bus Services in the British Isles,' *Geogr. Rev.*, **41**, 1951, 645–55 and many other papers.
13. GODLUND, S., articles in *Lund Studies in Geography*, nos. **17** and **18**, 1956 and also in ibid, no. **3**, 1951, with other papers.
14. These problems are discussed, in a neutral manner, in ref. 7.
15. ODUM, H. W. and MOORE, H. E., op. cit. in ref. 10, 29.
16. ibid. 110.
17. ibid. 168.
18. ibid. 188.
19. ibid. 195.
20. ibid. 248.
21. ibid. 217.
22. ibid. 222.
23. ibid. 219–20.
24. The work of Sten de Geer was first noted in *Ymer*, Stockholm, 1908, 240–53, with map 1:300,000, 'Befolkningens fördelning på Gottland': the atlas and text was published in 1919 as *Karta over Belfolkningens Fördelning i Sverige den 1 Januari*, 1917 (maps 1 to 500,000, 12 plates). See also *Geogr. Rev.*, **12**, 1922, 72–83; **13**, 1923, 497–506.
25. NEWBIGIN, M. I., *Mediterranean Lands*, London, 1924, chapter 4.
26. WHITTLESEY, D., 'Major Agricultural Regions of the Earth,' *Ann. Ass. Amer. Geogr.*, **26**, 1936, 199–240, esp. 202. This article has interesting comments on many previous works.
27. The numerous papers in *Economic Geography* on agriculture are of great significance. See ref. 35, and also SHANTZ, H. L., and MARBUT, C. F., *The Vegetation and Soils of Africa*, New York, 1923.
28. HARTSHORNE, R. *and* DICKEN, S. N., 'A Classification of the Agricultural Regions of Europe and North America on a Uniform Statistical Basis,' *Ann. Ass. Amer. Geogr.*, **15**, 1935, 99–120.
29. op. cit. in ref. 26, 108–40.
30. STAMP, L. D., *Our Undeveloped World*, London, 1953, esp. 101–04. But reference to pages is misleading, for the book should be read as a whole.
31. HALL, R. B., 'The Geographic Region: A Résumé,' *Ann. Ass. Amer. Geogr.*, **25**, 1935, 122–30, esp. 130.
32. *Econ. Geogr.*, **27**, 1951, 207.

33. op. cit. ref. 26.
34. op. cit. ref. 28.
35. BAKER, O. E., *Econ. Geogr.*, 2, 1926, 459–93; 3, 1927, 50–86, 309–39, 447–65; 4, 1928, 44–73, 399–433; 5, 1929, 36–69; 6, 1930, 166–90, 278–308; 7, 1931, 109–53, 325–64; 8, 1932, 325–77; 9, 1933, 167–97.
36. KENDREW, W. G., *Climates of the Continents*, 3rd. ed., 1937, 288.
37. DRYER, C. R., 'Natural Economic Regions,' *Ann. Ass. Amer. Geogr.*, 5, 1915, 122.
38. This is one theme of the *Tableau de la Géographie de la France*, Paris, 1903, and of several maps in the *Atlas Vidal-Lablache*. The full title is *Histoire et Géographie. Atlas général*, with editions in 1894, 1908, 1918, 1923, 1952.
39. In OGILVIE, A. G. (ed.) *Great Britain: Essays in Regional Geography*, Cambridge, 1928, 19–41.
40. ibid. 40.
41. ibid. 142–66.
42. ROXBY, P. M., 'The Agricultural Geography of England on a Regional Basis,' *Geogr. Teach.*, 7, 1913–14, 316–21.
43. This section is based partly on personal knowledge, but see his article noted in ref. 44, and 'The Scope and Aims of Human Geography,' *Scot. Geogr. Mag.*, 46, 1930, 276–90.
44. ROXBY, P. M., *Geogr. Teach.*, 13, 1925–6, 376–82.
45. STAMP, L. D., *The Land of Britain: Its Use and Misuse*, 2nd edn., 1950, deals especially with future planning in the last two chapters.
46. COLBY, C. C., 'Changing Currents of Geographic Thought in America,' *Ann. Ass. Amer. Geogr.*, 26, 1936, 35. But it could be urged that much has been added since 1936.
47. Published as *Fennia*, 72, Helsinki, 1952, by the Geographical Society of Finland, with J. G. GRANÖ as Chairman of the editorial board.
48. ibid. chapter by J. G. GRANÖ on geographic regions, 408–38.
49. ibid. 340–1.
50. MEURMAN, O.–I., in op. cit. 445–6.
51. This print may be seen by comparing such a work as the *Géogr. Univ.* volume on Russia by P. CAMENA D'ALMEIDA, 1932, with modern textbooks such as BALZAK, S. S., and others, *Economic Geography of the U.S.S.R.* ed. C. D. HARRIS, New York 1949; or the shorter Baransky, N., Moscow 1956 (same title); but a modern classic, BERG, L. S., *Natural Regions of the U.S.S.R.*, London, 1950, is on physical geography including climate. At present, the economic changes in the U.S.S.R. are so swift that students of the U.S.S.R. depend largely on periodicals: in English, a useful summary is given in *Soviet Geography: Review*

and Translation, ed. T. SHABAD, published monthly by the American Geographical Society.

52. UNSTEAD, J. F., 'A System of Regional Geography,' *Geography*, **27**, 1933, 178.

53. CHISHOLM, G. G., *Scot. Geogr. Mag.*, **24**, 1908, 575.

54. VIDAL DE LA BLACHE, *Principles of Human Geography*, trans. M. T. BINGHAM, London, 1926, is a translation of the French edition edited by E. DE. MARTONNE.

55. GARNETT, A., 'Insolation, Topography and Settlement in the Alps,' *Geogr. Rev.*, **25**, 1935, 601–17 and, a more extended treatment, 'Insolation and Relief—their Bearing on the Human Geography of Alpine Regions,' *Pub. Inst. Brit. Geogr.*, **5**, London, 1937.

56. Quoted in COLBY, C. C., in *Ann. Ass. Amer. Geogr.*, **26**, 1936, 8: the quotation is from the *Report on the United States and Mexican Boundary Survey*, Washington, 1857.

57. 'Two landscapes: South-west Scotland and North-east Ireland,' in *Geographical Essays in Memory of Alan G. Ogilvie*, ed. MILLER, R. and WATSON, J. W., London, 1959, 46–67.

58. Published 1942, with a revised edition in 1950.

CHAPTER SEVEN

1. WOOLDRIDGE, S. W. and EAST, W. G., *The Spirit and Purpose of Geography*, London, 1951, 103.

2. This work began in the late 1950's and a conference was held in London in December, 1959: regional councils have been formed in various parts of Britain.

3. CHISHOLM, G. G., *Handbook of Commercial Geography*, first sentence of his text.

4. ibid. 1.

5. CLAPHAM, J., *Economic History of Modern Britain*, Cambridge, 1926, I, 536.

6. LORIMER, F., *The Population of the Soviet Union: History and Prospects*, Geneva, 1946, 99.

7. FINCH, V. C., 'Training for Research in Economic Geography,' *Ann. Ass. Amer. Geogr.*, **34**, 1944, 214.

8. For an example, see JOHNSON, W. A. D., 'The Virgin and Idle Lands of Western Siberia and Northern Kazakhstan: a geographical appraisal,' *Geogr. Rev.*, **46**, 1956, 1–19. An estimate of American losses is given in WIBBERLEY, G. P., *Agriculture and Urban Growth*, London, 1959; this work includes an excellent summary of the problems of loss of land from farming in relation to increased productivity by modern agricultural methods.

9. JOHNSON, W. A. D., op. cit. in ref. 8, esp. 11–14.

10. STAMP, J. C., 'Geography and Economic Theory,' *Geography*, **22,** 1937, 1–14.
11. ibid. 9.
12. SIMPSON, E. S., 'The Cheshire Grass-Dairying Region,' *Pub. Inst. Brit. Geogr.*, **23,** 1957, 141–62.
13. CHISHOLM, G. G. 'The Goal of Commerce,' *Geogr. Teach.*, **12,** 1923–4, 333.
14. ibid. 338.
15. CHISHOLM, G. G. in *Scot. Geogr. Mag.*, **24,** 1908, 127.
16. McCARTY, H. H., 'An Approach to a Theory of Economic Geography,' *Econ. Geogr.*, **30,** 1954, 95–101.
17. In the case of Brazil, JAMES, P. E., *Latin America*, London, 1950, 521.
18. F.A.O., *Yearbook of Food and Agricultural Statistics* (annual).
19. Indian Famine Commission, *Report*.
20. WIBBERLEY, op. cit. in note 8, 220.
21. op. cit. 202–14 (on Lymm, Cheshire, a suggested area for re-housing from Manchester).
22. RUSSELL, J., *World Population and World Food Supplies*, London, 1954, 6.
23. ibid. 7.
24. BRUNHES, J., *Human Geography*, London, 1952, 156.
25. Noted in op. cit. 160.
26. *Econ. Geogr.* 27, 1951, 33–42.
27. MEAD, W. R., *An Economic Geography of the Scandinavian States and Finland*, London, 1958, 263.
28. HELM, E., 'A Review of the Cotton Trade of the United Kingdom, During the Seven Years, 1862–68,' *Transactions of the Manchester Statistical Society*, 1868–9, 67–94; for a later time, GREENWOOD, A., 'The Growth of the Cotton Trade,' *J. Manch. Geogr. Soc.*, **3,** 1887, 42–52.
29. HUTTON, W., *History of Birmingham*, 1781, quoted in GILL, C. *History of Birmingham*, Oxford, 1952, vol. I, 66.
30. LÖSCH, A., *The Economics of Location*, New Haven, 1954, 376–7.
31. ibid. 170, 490.
32. This is mentioned in ibid. 263.
33. ibid. 416.
34. ibid. xi.
35. BIRD, J., *The Geography of the Port of London*, London, 1957.
36. KANE, R., *The Industrial Resources of Ireland*, Dublin, 1844.
37. MARBUT, C. F., 'The Rise, Decline and Revival of Malthusianism in Relation to Geography and Character of Soils,' *Ann. Ass. Amer. Geogr.*, **15,** 1925, 1–28.
38. BAKER, O. E., in 'Conditions Determining Land Utilization,' *Ann. Ass. Amer. Geogr.*, **11,** 1921, 45–6.

39. ibid. 34–5.
40. ibid. 39.
41. CURRY, L., 'Climate and Economic Life,' *Geogr. Rev.*, **42**, 1952, 367–83. The word 'evapotranspiration' is taken from THORNTHWAITE, C. W., 'An Approach Toward a Rational Classification of Climate,' *Geogr. Rev.*, **38**, 1948, 55–94.
42. RUSSELL, J., *Geogr. Teach.*, **12**, 1923–4, 10–13.
43. ibid. 13.
44. SAUER, C. O., 'The Problems of Land Classification,' *Ann. Ass. Amer. Geogr.*, **11**, 1921, 3–16.
45. STAMP, L. D., *The Land of Britain: Its Use and Misuse*, 1948, 1950.
46. DRURY, G., *The Channel Islands*, London, 1950: in 'The Land of Britain, Report of the Land Utilization Survey of Britain,' ed. L. D. STAMP.
47. This is due largely to the enterprise of Miss A. C. Coleman, of King's College, London, and a few others.
48. See STAMP, L. D., *Applied Geography*, London, 1960, 46–50. The possible modification of the original scheme was discussed at the International Geographical Congress in Stockholm, 1960.
49. A summary of progress by 1960 is given in the Internation Geographical Union Newsletter, **11**, 1960, 38–46. Four volumes have been published (inquiries to Geographical Publications, Bude, Cornwall): COLE, MONICA M., *Land-use Studies in the Transvaal Lowveld*, 1956; TREGEAR, T. R., *Land-use in Hong Kong and the New Territories*, 1958; CHRISTODOULOU, D., *The Evolution of the Rural Land-use Pattern in Cyprus*, 1960; NIDDRIE, D. L., *Land-use and Population in Tobago*, 1961.
50. STAMP, L. D., *Our Undeveloped World*, 1953.
51. *Census of Ireland*, 1851, xvii.

CHAPTER EIGHT

1. Neither work was intended to be a complete statement on human geography: VIDAL DE LA BLACHE'S book was not finished at the time of his death, and was moulded by E. DE MARTONNE from the material found. BRUNHES'S book includes many 'samples'; in the 1952 abridged English edition, it is noted (p. 6) that 'his primary concern was to introduce a few principles of "geographical logic" in the fragmentary sketches in the economic chapters.' J. BRUNHES, *La Géographie Humaine*, was first published in 1910, and a revised and enlarged edition was issued in 1912: the fourth edition is dated 1934. The English translation,

New York, 1920, was by T. C. LE COMPTE, and edited by I. BOWMAN and R. E. DODGE: the abridged edition was by Mme. M. Jean-Brunhes Delamarre and P. Deffontaines, trans. E. F. Rowe, Paris, 1947, London, 1952.

2. PEATTIE, R., *Mountain Geography*, Cambridge, Mass., 1936.
3. HARTSHORNE, R., *The Nature of Geography*, 1939, 76.
4. VIDAL DE LA BLACHE, *Principles of Human Geography*, 1926, 6.
5. ibid. 9.
6. ibid. 10.
7. ibid. 14–15.
8. ibid. 153.
9. This 'dualism' of geography has been discussed recently by several authors: a recent statement is HARTSHORNE, R., *Perspective on the Nature of Geography*, Chicago and London, 1959, esp. 65–80, the chapter on 'the division of geography by topical fields—the dualism of physical and human geography.'
10. BRUNHES, J., *Human Geography*, London, 1952, 227.
11. In his novel, *Coningsby*.
12. CLARK, J. G. D., *The Mesolithic Settlement of Northern Europe*, Cambridge, 1936, and the earlier work on Britain, 1932. The 1936 book includes the statement (p. xiii), 'The great influence exercised by physical environment has for a long time been a commonplace of archaeological and anthropological research; it is less generally recognized that this environment has undergone changes in the last few thousand years so profound as to alter its influence on cultural development and so rapid as to afford a natural time-scale for the dating and synchronizing of human cultures.'
13. LEBON, J. H. G., *An Introduction to Human Geography*, London, 2nd edn., 1959.
14. ROXBY, P. M., 'The Scope and Aims of Human Geography,' *Scot. Geogr. Mag.*, **46,** 1930, 276–99.
15. FLEURE, H. J., *An Introduction to Geography*, London, 1929, 75.
16. FLEURE, H. J., *The Races of Mankind*, London, 1927, 70.
17. ibid. 68.
18. ibid. 77.
19. PENNIMAN, T. K., *A Hundred Years of Anthropology*, 1935, 1952. There were nine books (Oxford) when PENNIMAN wrote, published from 1927 (*Apes and Men*) to 1936 (*The Horse and the Sword*), but later, in 1956, vol. X, *Times and Places*, appeared.
20. This, perhaps the most frequently quoted of all geographical phrases, appears at the end of an article by P. VIDAL DE LA BLACHE, 'Des Caractères Distinctifs de la Géographie,' *Ann. Géogr.*, **22,** 1913, 289–99.
21. FLEURE, H. J., op. cit. in ref. 15, 13. Like many liberal thinkers

of his time, FLEURE was strongly attracted to the thought of the great South African statesman.

22. But this is not a universal view: see, for example, FLEURE, H. J., 'Geographical Thought in the Changing World,' *Geogr. Rev.*, **43**, 1944, 515–28.

23. TREWARTHA, G. T., 'A Case for Population Geography,' *Ann. Ass. Amer. Geogr.*, **43**, 1953, 71–97.

24. DICKINSON, R. E., *City, Region and Regionalism*, London, 1947, xiii: however the next two sentences read, 'But it would be futile and sheer frustration to circumscribe study in this field, as in so many other problems of contemporary society, by the arbitrarily fixed limits of particular problems. What matters is the problem.'

25. TREWARTHA, G. T., op. cit. 83.

26. The whole of Britain is covered by county reports, of which a complete list is given in STAMP, L. D., *The Land of Britain: Its Use and Misuse*, 2nd edn. 1950. In a few cases, more than one county is covered in one volume.

27. The European Society for Rural Sociology has already provided a general forum for students of such problems, at its congresses in Louvain, 1958, and Norway, 1960, and in its journal *Sociologia Ruralis*.

28. TREWARTHA, G. T., op. cit. 95.

29. FREEMAN, T. W., 'The Congested District of Western Ireland,' *Geogr. Rev.*, **33**, 1943, 1–14.

30. TREWARTHA, G. T., op. cit. 94.

31. VINCE, S. W. E., 'Reflections on the Structure and Distribution of Rural Population in England and Wales, 1921–31.' *Pub. Inst. Brit. Geogr.*, **18**, 1953, 53–76.

32. STEVENS, A., 'The Distribution of the Rural Population of Great Britain,' *Pub. Inst. Brit. Geogr.*, **11**, 1946, 21–54.

33. FREEMAN, T. W., *Ireland, a General and Regional Geography*, 2nd. Ed., London, 1960, 132–4.

34. *Geogr. J.*, **89**, 1937, 362.

35. *Lund Studies in Geography, Ser. B. Human Geography*, no. **13**, 'Migration in Sweden, A Symposium,' ed. D. HANNERBERG and others, Lund, 1957.

36. ROXBY, P. M., 'The Distribution of Population in China,' *Geogr. Rev.*, **15**, 1–24.

37. KING, F. H., *Farmers of Forty Centuries*, Madison, Wis., 1911; republished, edited by J. P. BRUCE, London, 1927, 1933, 1939, 1949. See also MALLORY, W. H., *China: Land of Famine*, New York, 1926.

38. ROXBY, P. M., op. cit. 16–17.

39. Quoted in *Geogr. J.*, **106**, 1945, 123.

40. GEDDES, A., 'Half a Century of Population Trends in India: A Regional Study of Net Change and Variability, 1881–1931,' *Geogr. J.*, **98**, 1941, 228–53, 'The Population of India: Variability of Change as a Regional Demographic Index,' *Geogr. Rev.*, **32**, 1942, 562–73.

41. GILLMAN, C., 'A Population Map of Tanganyika Territory,' *Geogr. Rev.*, **26**, 1936, 353–75.

42. TREWARTHA, G. T., op. cit. in ref. 23, 88–95.

43. *American Geography, Inventory and Prospect*, 108–09.

44. This famous survey includes a vast amount of social information.

45. *American Geography*, op. cit. ref. 43, 159.

46. Books on urban geography include CHABOT, G., *Les Villes*, Paris, 1948; DICKINSON, R. E., *City, Region and Regionalism*, London, 1947, *The West European City*, London, 1951; GEORGE, P., *La Ville: le Fait Urbain à Travers le Monde*, Paris, 1954; SMAILES, A. E., *The Geography of Towns*, London, 1953; SORRE, M., *Les Fondements de la Géographie Humaine*, tome 3, *L'habitat*, Paris, 1952; TAYLOR, G., *Urban Geography*, London, 1949. See the bibliography (Erdkunde) noted in ref. 49.

47. BLANCHARD, R., *Grenoble*, Paris, 1911 deals with the river, the site, history and finally has chapters on Grenoble as an industrial town and 'capitale régionale'. LEIGHLY, J. B., 'The Towns of Malardalen in Sweden: A Study in Urban Morphology,' *University of California Publications in Geography*, **3**, 1928–30, Berkeley, Calif. 1–135.

48. These handbooks do not receive, in some years, the circulation they deserve, for many of them contain a great deal of valuable and previously unpublished material.

49. The bibliography is too extensive to list here, but some recent works are noted in *American Geography Inventory and Prospect*, 162–6, also in *La Géographie Française au Milieu du XX Siècle* (in an article by G. CHABOT) and in DICKINSON, R. E., op. cit. in ref. 24. An excellent bibliography (427 items) is given in *Erdkunde*, **7**, Bonn, 1953, 161–84.

50. HARRIS, C. D., 'A Functional Classification of Cities in the United States,' *Geogr. Rev.*, **33**, 1943, 86–99.

51. ADSHEAD'S *Twenty-Four Illustrated Maps of the Township of Manchester Divided into Municipal Wards Corrected to the 1st May, 1851*, published by Joseph Adshead, no. 45 George St., Manchester. The original survey was 'by Richard Thornton, carefully corrected April 24th 1850'. There is a copy in the map room of the British Museum (Box 3215, Map 18), and in the Central Reference library, Manchester.

52. More will be heard of industrial archaeology in the near future.

53. This information, and a vast amount more, is available in the marginal notes of the nineteenth-century Censuses.

54. GILBERT, E. W., 'Pioneer Maps of Health and Disease in England,' *Geogr. J.*, **124**, 1958, 172–83.

55. 'Report from the Select Committee on the Health of Towns.' *House of Commons Papers, Reports from Committees* vol. 8, no. 384, xv–xxii.

56. The phrase 'neighbourhood unit' has become conventional: it is discussed in PERRY, C. A., *Housing for the Machine Age*, New York, 1939.

57. A vivid treatment of the theme is given in SELF, P., *Cities in Flood*, London, 1957.

58. WILLIAM-OLSSON, W., 'Stockholm: Its Structure and Development,' *Geogr. Rev.*, **30**, 1940, 420–38: WILLIAM-OLSSON also issued a study of Stockholm in 1960 for the International Congress, *Stockholm: Structure and Development*: BENYON, E. D., 'Budapest: An Ecological Study,' *Geogr. Rev.*, **33**, 1943, 256–75.

59. CLOZIER, R., *La Gare du Nord*, Paris, 1941.

60. This work of E. W. BURGESS appeared in PARK, R. E., *The City*, Chicago, 1923.

61. See op. cit. in ref. 55, 21.

62. These are thoroughly discussed in DICKINSON, R. E., *City, Region and Regionalism*, 30–5, 45, 53–62, 90–1.

63. ibid. 57.

64. *Census of Great Britain*, 1851, vol. I, xiv-xlvii. See also ref. 55.

65. SMAILES, A. E., 'The Urban Hierarchy in England and Wales,' *Geography*, **29**, 1944, 41–51; 'The Urban Mesh of England and Wales,' *Pub. Inst. Brit. Geogr.*, **11**, 1946, 87–101.

66. *American Geography, Inventory and Prospect*, ref. 6, chapter one, 144.

67. BRACEY, H. E., *Social Provision in Rural Wiltshire*, London, 1952: 'Towns as Rural Service Centres,' *Pub. Inst. Brit. Geogr.*, **19**, 1953, 95–105; 'Rural Component of Centrality Applied to Six Southern Counties in the United Kingdom,' *Econ. Geogr.*, **32**, 1956, 38–50, and ref. 65.

68. GREEN, F. H. W., 'Urban Hinterlands in England and Wales: An Analysis of Bus Services,' *Geogr. J.*, **116**, 1950, 64–88; 'Bus Services in the British Isles,' *Geogr. Rev.*, **41**, 1951, 645–55 and many other papers.

69. GEDDES, P., *Cities in Evolution*, London, 1915.

70. FAWCETT, C. B., 'Distribution of the Urban Population in Great Britain,' *Geogr. J.*, **79**, 1932, 100–13. A modern treatment is FREEMAN, T. W., *The Conurbations of Great Britain*, Manchester, 1959.

71. FAWCETT, C. B., *The Provinces of England*, originally published in 1919, was re-issued with some revisions by W. G. EAST and S. W. WOOLDRIDGE in 1960.

72. *American Geography: Inventory and Prospect*, chapter on medical geography 453–68 and articles in *Geogr. Rev.*

CHAPTER NINE

1. OGILVIE, A. G., 'Isaiah Bowman: An Appreciation,' *Geogr. J.*, **115**, 1950.

2. Information from an anonymous pamphlet, Royal Geographical Society, no date, *Note sur les Travaux Scientifiques de M. Jovan Cvijić*.

3. CVIJIĆ, J., *La Peninsule Balkanique*, 1918, 2.

4. FREEMAN, E. A., *Historical Geography of Europe*, 1881, vi.

5. This is partly personal observation at the Amsterdam congress of 1938.

6. In *American Geography: Inventory and Prospect*, 174.

7. MARTIN, L., 'The Geography of the Monroe Doctrine and the Limits of the Western Hemisphere,' *Geogr. Rev.*, **30**, 1940, 525–8.

8. JAMES, P. *Latin America*, 1950.

9. WANKLYN, H. G. (Mrs J. A. Steers), 'The Middle People: Resettlement in Czechoslovakia,' *Geogr. Rev.*, **112**, 1948, 28–42: and *Czechoslovakia*, London, 1954.

10. For the inter-war period I. BOWMAN, *New World*, 1921, and later editions, remains a classic.

11. HARTSHORNE, R., 'The Functional Approach in Political Geography,' *Ann. Ass. Amer. Geogr.*, **40**, 150, esp. 116: the idea of the core area is discussed in OGILVIE, A. G., *Europe and Its Borderlands*, Edinburgh, 1957, 252–61.

12. ROXBY, P. M., 'The Scope and Aims of Human Geography,' *Scot. geogr. Mag.*, **46**, 1930, 287.

13. In *American Geography: Inventory and Prospect*, 177–8.

14. DOMINIAN, L., *The Frontiers of Language and Nationality in Europe*, New York, 1917, 315.

15. ibid. 316.

16. SCHRADER, F., 'The Foundations of Geography in the XXth Century,' *Geogr. Teach.*, **10**, 1919–20, 44–53, esp. 52: see, for a careful criticism of such ideas, FEBVRE, L., *A Geographical Introduction to History*, London and New York, 1925, 108–10, 117, 308–9, 334–7: Febvre's book was originally published as *La Terre et l'Evolution Humaine: Introduction géographique à l'Histoire*, 1922, and translated by E. G. MOUNTFORD and J. H. PAXTON.

17. Dominian, op. cit. 315–16, 329.
18. ibid. 185.
19. ibid. vii, 323–5.
20. ibid. 314.
21. See ref. 11.
22. Dominian, op. cit. 330.
23. ibid. 46–7.
24. VIDAL DE LA BLACHE, *La France de l'Est*, Paris, 1919 (3rd edn.), esp. 223–32.
25. Several of the plebiscite areas are excellently treated in I. BOWMAN, *New World*, edition of 1928: for example, Upper Silesia, 412–15.
26. See ref. 2.
27. CVIJIĆ, J., *La Péninsule Balkanique*, Paris, 1918, ii. A paper of historical interest is Cvijić's *L'annexion de la Bosnia et la Question Serbe*, Paris, 1909. This includes a map of nationalities, with an overprint showing the distribution of Roman Catholics.
28. CVIJIĆ, J., *La Peninsule Balkanique*, Paris, 1918, 6–7.
29. ibid. 268.
30. ibid. esp. 177–84, and map opposite 170.
31. ibid. 100–11: on climate, 36–43, natural regions, in part II of the book, 45–79.
32. These views are strongly stated in a review by J. CVIJIĆ of a book by A. ISCHIRKOV, *Les Confins Occidentaux des Terres Bulgares*, Lausanne, 1906, in *Bulletin de la Societé Neuchâteloise de Géographie*, **25**, 1916, 166–80.
33. CVIJIĆ, op. cit. in ref. 28, 98–9.
34. ibid. 6–7.
35. ibid. 508.
36. ibid. 509.
37. Commander Roncagli, of the Royal Italian Navy, was secretary of the Geographical Society at Rome. His paper, 'Physical and Strategical Geography of the Adriatic,' appeared in *Geogr. J.*, **53**, 1919, 209–28.
38. MARINELLI, O., 'The Regions of Mixed Populations in Northern Italy,' *Geogr. Rev.*, **7**, 1919, 129–48.
39. LOCKY, L. (ed.) *A Geographical, Economic and Social Survey of Hungary*, Budapest, 'Patria' Press, 1919.
40. The bibliographies in Bowman's *New World* are valuable.
41. CZAPLICKA, M. A., in *Geogr. J.*, **53**, 1919, 361–81.
42. WANKLYN (Steers), H. G., *The Eastern Marchlands of Europe*, London, 1941, 147, 161.
43. ROMER, E., 'Statistics of Languages of the Provinces being under the Polish Civil administration of the Eastern Lands' *Geographical Works, Memoir* **7**, Lwow–Warsaw 1919, 15–23, esp. 16: see also LUTOSTAWSKI, W., and ROMER, E., *The Ruthenian*

Question in Galicia, Paris, 1919, 3, 9–10. Another paper by E. ROMER is 'Problèmes Territoriaux de la Pologne,' *Scientia,* **28,** Bologna, 1920, 3–16.

44. GOTTMAN, J., in obit. of E. DE MARTONNE, *Geogr. Rev.,* **46,** 1956, 277–9. The extent to which the geographers influenced the Treaty of Versailles and subsequent arrangements of boundaries is doubtful, but may ultimately be known from the private papers of I. BOWMAN (in *Geogr. Rev.,* **41,** 1951, 51, it is noted that these were presented on his death to John Hopkins University but to remain sealed for twenty-five years). D. W. JOHNSON, in *Natural History,* **19,** New York, 1919, 516, says that 'the advice of the territorial experts was frequently sought and extensively used and . . . played no inconsiderable role in establishing the new frontiers of Europe'. BOWMAN, in *What Really Happened at Paris,* ed. E. M. HOUSE and C. SEYMOUR, London and New York, 1921, on pp. 161–2 describes the drawing of the boundaries of the Free City of Danzig: in the same book, 457–8, D. W. JOHNSON comments on the drawing of the boundary through the Istrian Peninsula. In H. W. V. TEMPERLEY (ed.) *History of the Paris Peace Conference,* London, 1921, vol. 4, 207–12, it is noted that a strong case was made for the creation of Yugoslavia which 'bore obvious traces of the hand of M. CVIJIĆ, the most learned and most enlightened not only of Serbian, but of all Balkan geographic experts,' There are some curious but revealing pamphlets of the time: one in the Royal Geographical Society, 'What I have seen in Besserabia,' by E. DE. MARTONNE, was a translation of an article in the *Revue de Paris,* 1 November, 1919. Vividly written travel rather than academic geography, on p. 46 it reads 'Besserabia . . . is a land of marvellous wealth, a possession of priceless value for Romania; it has a very mixed population in the south and all the towns, a purely Russian one in the centre, an illiterate peasantry, a Russified middle class. But just let things follow their normal course. The country is preparing to throw in its lot definitely with Romania.'

45. JOHNSON, D. W., 'A Geographer at the Front and at the Peace Conference,' *Natural History,* **19,** 1919, 511–21. This paper consists of a brief but excellent review of the strategy of the 1914–18 war, and on pp. 515–16 an account of the mechanism of finding new frontiers; 'the frontiers of the New Europe as you will see them on the map were, for the most part, drawn in the territorial commissions by disinterested geographic and other experts.' See also Johnson's *Topography and Strategy in the War,* New York, 1917.

46. WRIGLEY, G. M., in *Geogr. Rev.,* **41,** 1951, 21–23.

47. Bowman, I., in *What Really Happened at Paris*, ed. E. M. House and C. Seymour, London, 1921, 161-2.

48. Meyer, H. C. 'Mitteleuropa in German Political Geography,' *Ann. Ass. Amer. Geogr.*, **36**, 1946, 178-94.

49. Ormsby, H., 'The Definition of Mitteleuropa and Its Relation to the Conception of Deutschland in the Writings of Modern German Geographers.' *Scot. Geogr. Mag.*, **51**, 1935, 337-47 attracted attention when given at the 1935 British Association meeting: in the subsequent discussion one geographer attacked Mrs Ormsby for condoning the Treaty of Versailles, to which she gently retorted that her paper was not concerned with the treaty at all. A modern paper, with a fine bibliography, is Sinnhuber, K. A., 'Central Europe—Mitteleuropa—Europe Centrale,' *Pub. Inst. Brit. Geogr.*, **20**, 1954, 15-39. Some of the definitions appear to have been made by negative means—that is, by defining western, northern, eastern and Mediterranean Europe, and calling the residue 'central'. Far more interesting is the equation of central Europe with 'cultural' features.

50. Partsch, J., *Central Europe*, Cambridge, 1903, 142.

51. Sinnhuber, op. cit. in ref. 45, 23; 29.

52. op. cit. in ref. 37, 211-15.

53. Fischer, E., 'German Geographical Literature,' *Geogr. Rev.*, **36**, 1946, 92-100.

54. The German translation *Der Staat als Lebensform*, appeared in 1917.

55. Kiss, G., 'Political Geography into Geopolitics,' *Geogr. Rev.*, **32**, 1942, 632-45.

56. Mauco, G., 'La Situation Demographique de la France,' *Ann. Géogr.*, **48**, 1939, 85-90.

CHAPTER TEN

1. Freeman, E. A., *The Historical Geography of Europe*, London, 1881, included an atlas volumes with sixty-five maps by E. Weller.

2. Readers of Mrs Gaskell's *Life of Charlotte Brontë* will remember that when the Brontë sisters tried to establish a school they included 'the use of the globes'.

3. Among books with bibliographies are Hinks, A. R., *Maps and Survey*, Cambridge, 1913, and later editions (5th., 1947); Lynam, E., *British Maps and Map-makers*, London, 1944; Skelton, R. A., *Decorative Printed Maps of the Fifteenth to Eighteenth Centuries*, London, 1952; Crone, G. R., *Maps and Their Makers*, London, 1953. A classic bibliography is Chubb,

T., *The Printed Maps in the Atlases of Great Britain and Ireland*, London, 1927.

4. ROBINSON, A. H. (and others) in the article on 'Geographic Cartography', with an excellent bibliography, *American Geography: Inventory and Prospect*, 1954.

5. *Geogr. J.*, **5**, 1895, 369.

6. PETERMANN, A., *Atlas of Physical Geography*, London, 1850, in preface (no page number).

7. Available in the map room of the British Museum.

8. See vol. 1, 11.

9. GILBERT, E. W., 'Pioneer Maps of Health and Disease in England,' *Geogr. J.*, **124**, 1958, 172–83.

10. ROBINSON, A. H., 'The 1837 Maps of Henry Drury Harness,' *Geogr. J.*, **121**, 1955, 440–50; and note in *Geogr. J.*, **122**, 1956, 129–31.

11. The full title of the Census is *Report of the Commissioners appointed to take the Census of Ireland for the Year* 1841, Dublin, 1843. The general maps appear at the end of the introduction, and the Dublin map between pp. lxxiv and lxxv in the 'Report upon the table of deaths'.

12. This survey was published in nine volumes in 1892–97: a revised and enlarged edition was published in 1902–04.

13. Published by Joseph Adshead, 45 George Street, Manchester. See ref. 51, chapter 8.

14. Report from the Select Committee on the Health of Towns, *House of Commons Paper* 1840, *Reports from Committees*, vol. 8, no. 384, xv–xxii.

15. Published London, 1831: the third edition is dated 1837. See preface to vol. 5, p. ii.

16. op. cit. in note 10, 449.

17. PETERMANN, A., *Map of the British Isles Elucidating the Distribution of the Population of the British Isles Based on the Census of* 1841, London, 1849. There are copies of this work in the map room of the British Museum, and in the Department of Geography, Trinity College, Dublin.

18. Apart from the excellent maps mentioned, there are several others of considerable interest, such as maps showing the sizes of towns by various symbols, the registration counties and districts, the registration districts of London and, for parts of Essex, the extremely intricate arrangements of parishes.

19. ROBINSON, op. cit. in ref. 10, 449.

20. ibid. 449.

21. Census of Ireland, 1881, *General Report*, 77.

22. *J. R. Geogr. Soc.*, **50**, 1879–80, 94.

24. ibid. 116–17.

25. ibid. 116.

26. op. cit. in ref. 23, 702–04.

27. HOSIE, A., 'A Journey in South-western China, from Ssu-ch'uan to Western Yunnan,' *Proc. R. Geogr. Soc.*, **8**, 1886, 371–84: *Three Years in Western China: A Narrative of Three Journeys in Ssu-ch'uan, Kuei-chow, and Yun-nan*, with an introduction by A. LITTLE, London, 1890.

28. Both the RITTER maps of 1806 and the HUMBOLDT 1812 atlas are in the Royal Geographical Society, London.

29. The first edition appeared in 1834; the ninth in 1905 (with, significantly, revisions in 1907, 1909, 1914, 1915, 1916) and the tenth in 1930.

30. HINKS, op. cit. in note 3 (3rd ed., 1933, 118) grimly notes that ANDREE'S *Allgemeiner Handatlas*, 8th Ed., Bielefeld und Leipzig, weighs 14 lbs. and STIELER'S Hand-Atlas, 1930, 12 lbs.

31. Published for the Society for the Propagation of the Gospel, London; printed by R. Clay, Bread Street, Hull. The maps were 'drawn and engraved by J. ARCHER'.

32. BERGHAUS, H., *Physikalischer Atlas*, published at Gotha.

33. The full title is 'The national atlas of historical, commercial and political geography constructed from the most recent and authentic sources by Alexander Keith Johnston, Geographer at Edinburgh in ordinary to Her Majesty, accompanied by maps and illustrations of the physical geography of the globe by Dr Heinrich Berghaus, professor of Geography, Berlin, and an ethnographic map of Europe by Dr Gustav Kombst.'

34. JOHNSTON, A. K., *Physical Atlas of Natural Phenomena*, Edinburgh and London, 1850, 1856.

35. op. cit. note 6.

36. The five volumes are I, *The World, Australasia and East Asia*, 1958; II, *South-west Asia and Russia*, 1959; III, *Northern Europe*, 1955; IV, *Southern Europe and Africa*, 1956; V, *The Americas*, 1957.

37. cf. Hinks, op. cit. in note 3, 117.

38. BARTHOLOMEW, J. G., *The Royal Scottish Geographical Society's Atlas of Scotland*, Edinburgh, 1895. Among the editors were A. GEIKIE for geology, J. GEIKIE on physiography and A. BUCHAN on meteorology. A revised edition appeared in 1912.

39. Bartholomew's *Physical Atlas*: vol. 3, *Atlas of Meteorology, a Series of Over Four Hundred Maps*, prepared by J. G. BARTHOLOMEW and A. J. HERBERTSON and edited by A. BUCHAN, Westminster, 1899: vol. 5, *Atlas of Zoogeography, a Series of Maps Illustrating the Distribution of over Seven Hundred Families, Genera and Species of Existing Animals*, prepared by J. G.

BARTHOLOMEW, W. E. CLARKE and P. H. GRIMSHAW, London, 1911. Vols 1, 2, and 4 never appeared.

40. This atlas was produced by the famous Armand Colin firm.

41. PAULLIN, C. O., ed. WRIGHT, J. K., *Atlas of the Historical Geography of the United States*, published jointly by the Carnegie Institute of Washington and the American Geographical Society of New York, 1932. The story of the compilation of this atlas is told in WRIGHT, J. K., *Geography in the Making*, New York, 1952, 291–3. It includes an introduction of 162 pages and an index.

42. A useful summary, with an excellent bibliography, is YONGE, E. L., 'National Atlases,' *Geogr. Rev.*, **47**, 1957, 570–8.

43. This appeared in *Atlas de Finlande*, (by) Société de Géographie de Finlande, Helsingfors, 1899.

44. The 1906 *Atlas of Canada* was published in Ottawa by the Department of the Interior and 'prepared under the direction of James White'. The 1915 Edition is described as 'revised and enlarged', and the 1957 Atlas of Canada was published in 1959 at Ottawa by the Geographical Branch, Department of Mines and Technical Surveys.

45. The *Atlas de France* is published by the Comité de Géographie, Paris, of which E. de Martonne was general secretary and E. de Margerie president of the Commission for the Atlas. There are four physical maps, 1933, and four morphological maps by E. de Martonne and others, 1938 and 1941.

46. *Atlas de Belgique*, Brussels, Comité Nationale de Géographie, 1951. The morphological map was the work of M. A. LEFÈVRE.

47. *Atlas de la République Tchécoslavaque*, publié par L'Académie Tchéque, Prague, 1935. The atlas is published in Czech and French.

CHAPTER ELEVEN

1. MILL, H. R., *An Autobiography*, London, 1951: TAYLOR, G., *Journeyman Taylor*, London, 1958.

2. LOWENTHAL, D., *George Perkins Marsh: Versatile Vermonter*, New York, 1958.

3. HARTSHORNE, R., *Perspective on the Nature of Geography*, New York, 1959, 60, 62–3.

4. Originally published in *Ann. Ass. Amer. Geogr.*, **29**, 1939, 173–658.

5. WHITTLESEY, D., 'Field Study of Human Geography in the United States,' *Publications of the International Congrés*, Paris, 1931, tome III, 774–8. Whittlesey describes how 'towards 1913' some efforts were made 'to systematize the study of human

geography' and 'in 1915 two young instructors published the first American statement of objectives in the field study of Human Geography'. See JONES, W. D., and SAUER, C. O., *Bull. Amer. Geogr. Soc.*, **47**, 1915, 520–5: from this came other papers including SAUER, C. O., in *Geogr. Rev.*, **8**, 1919, 47–54 and *Ann. Ass. Amer. Geogr.*, **14**, 1924, 17–33; and others in ibid. by W. D. JONES and V. C. FINCH in 1925, by D. WHITTLESEY, in 1925 and 1927 with a practical example by A. E. PARKINS on Nashville, ibid, **20**, 1930, 164–75.

6. TAYLOR, G. (ed.) *Geography in the Twentieth Century*, London, 1950, essay on 'Geopolitics and Geopacifics', 587–608.

7. The world land-use survey is a concern of the International Geographical Union: already (1961) some maps of the new survey of land-use in Britain have been published—inquiries to Dr Alice Coleman, King's College, Strand, London, W. C. 2.

8. This is given in *Erdkunde*, **7**, Bonn, 1953, 179–84: in all 427 references are given. The article is SCHOLLER, P., 'Aufgaben und Probleme de Stadtgeographie,' 161–84.

9. GOTTMAN, J., 'Megalopolis or the Urbanization of the North-eastern Seaboard,' *Econ. Geogr.*, **33**, 1957, 189–200.

10. WILLIAM-OLSSON, W., 'Stockholm Its Structure and Development,' *Geogr. Rev.*, **30**, 1940, 420–38 (a summary of a work in Swedish published 1937, in ref. 1 of article): *Stockholm: Structure and Development*, International Geographical Congress, 1960.

11. A famous pioneer study is BLANCHARD, R., *Grenoble: Étude de Géographie Urbaine*, Paris, 1911, 1912, 1935.

12. This view is expounded in DICKINSON, R. E., *West European City*, London, 1951.

13. BROWN, R. H., *Mirror for Americans: Likeness of the Eastern Seaboard, 1810*, New York, 1943.

14. An excellent early study is GOLDTHWAIT, J. W., 'A Town that has Gone Downhill,' *Geogr. Rev.*, **17**, 1927, 527–53.

15. BERESFORD, M., *History on the Ground*, London, 1957, esp. 95–100. That the Black Death was not necessarily responsible for the downfall of villages is shown in the same author's *Lost Villages of England*, London, 1954, esp. 158–66: see also BERESFORD, M. W. and ST JOSEPH, J. K. S., *Medieval England, An Aerial Survey*, Cambridge, 1958.

16. CRAWFORD, O. G. S., *Archaeology in the Field*, London, 1953, 45–8; KEILLER, A. and CRAWFORD, O. G. S., *Wessex from the Air*, Cambridge, 1928, esp. 3–7.

17. GILBERT, E. W., 'What is Historical Geography?' *Scot. Geogr. Mag.*, **48**, 1932, 129–36.

18. DARBY, H. C., (ed.), *An Historical Geography of England before*

A.D. 1800, Cambridge, 1936, and DARBY, H. C., 'An Historical Geography of England: Twenty Years After,' *Geogr. J.*, **126**, 1960, 147–59.

19. DARBY, H. C., *Domesday Geography: Eastern England*, Cambridge, 1952; *Midland England*, ed. H. C. DARBY and I. B. TERRETT, Cambridge, 1954.

20. There is, for example, a thorough survey in the Manchester Central Reference Library, and doubtless in many other libraries.

21. TAYLOR, G. op. cit. in ref. 1, 138–41, 168–74, 325–33, 345.

22. WRIGHT, J. K., *Geography in the Making*, New York, 1952, 300–19.

23. This enterprise is sponsored by the World Population Commission of the International Geographical Union.

24. First published in the fifth edition of the *Oxford Advanced Atlas*, by J. BARTHOLOMEW, Oxford, 1936, and later, in Bartholomew's *Advanced Atlas*, London, 1950.

25. HARDY, M. E., *The Geography of Plants*, Oxford, 1920.

26. VIDAL DE LA BLACHE, *Principles of Human Geography*, London, 1926, esp. 49–160, 211–37.

27. HARTSHORNE, R., *Perspective on the Nature of Geography*, Chicago and London, 1959, 20–1, 47, 172.

28. BOWEN, E. G., 'Welsh Emigration Overseas,' *Advanc. Sci.*, **17**, 1960, 265–70.

29. This was originally done for the Ministry of Town and Country Planning and privately published, but the problem of derelict land has received attention in various works. See for example West Midland Group, *Conurbation: A Planning Scheme of Birmingham and the Black Country*, London, 1948; BEAVER, S. H., 'Land Reclamation after Surface Mineral Working,' *Journal of the Town Planning Institute*, **41**, 1955, 146–54; WALLWORK, K., 'Subsidence in the Mid-Cheshire Area, *Geogr. J.*, **122**, 1596, 40–53, 'Some Problems of Subsidence and land-use in the Mid-Cheshire Industrial Area,' *Geogr. J.*, **126**, 1960, 191–9.

30. Quoted in HARTSHORNE, op. cit. in ref. 26, 62.

APPENDIX
SHORT BIOGRAPHIES OF GEOGRAPHERS

This section follows the example given by O. J. R. Howarth in *Advanc. Sci.*, **30**, 1951, 162–4: Dr Howarth included numerous geographers not mentioned here, particularly those connected with the British Association and the Royal Geographical Society. In the following list useful sources have been indicated for those who may wish to read further. The obituaries of the past are of varied quality: some fail to give either the year of birth or death: many do not give the full name and several include only the vaguest reference to the published works though others particularly for some of the major American and German geographers and in recent years for those of Great Britain (in *Pub. Inst. Brit. Geogr.*) give a full bibliography. Indeed the Institute of British Geographers asks its members to provide this information before demise.

'Ere you mount the heavenly stair
Detailed references prepare.

ARROWSMITH, JOHN. 1790–1873.
A cartographer with his uncle Aaron (ob. 1822), he used the records of explorers to compile maps of the continents, notably of Australia. Keenly interested in African discovery, he was a friend of Livingstone.
J. R. Geogr. Soc., **43**, 1873, clxi–clxiii.

ATWOOD, WALLACE WALTER. 1872–1949.
He was gradually drawn into geography from geology, and from 1901–13 served on the teaching staff at Chicago University with T. C. Chamberlain and R. D. Salisbury, both well known as physiographers. In 1913 he succeeded W. M. Davis at Harvard, and in 1920 he became president of Clark University where he founded the graduate school and the journal *Economic Geography*, long edited by W. Elmer Ekblaw. Atwood wrote largely on physiography but was concerned with wider geographical problems, and produced a number of textbooks and maps.
Ann. Ass. Amer. Geogr., **39**, 1949, 296–306. *Geogr. Rev.*, **39**, 1949, 675–7.

BAKER, OLIVER EDWIN. 1883–1950.
His graduate work on agriculture from 1908 set the pattern for his

303

lifelong devotion to farming; in 1912 he joined the U.S. Department of Agriculture, and contributed to the *Yearbooks* from 1915–38, some of which he edited. He was part-author of the Department's *Atlas of World Agriculture*, 1917 and compiled the *Atlas of American Agriculture*, published in parts from 1918–36. His works included numerous articles in *Economic Geography* and he became increasingly concerned with general problems of land use and population.

Ann. Ass. Amer. Geogr., **40**, 1950, 328–34: *Geogr. Rev.*, **40**, 1950, 333–4.

BARTHOLOMEW, JOHN GEORGE. 1860–1920.

The elder son of John Bartholomew, founder of the map firm, he entered the draughtsman's office at seventeen and experimented on the replacement of hill-shading by contours and layer colouring. The first maps with this technique appeared in Baddeley's *Guide to the Lake District*, 1880, and the half inch to one mile maps of the British Isles, with layer colouring, were published from 1888. He helped Sir John Murray with the maps in the 'Challenger' *Reports* and Dr Buchan with the *Meteorological Atlas* published by Bartholomew.

Scot. Geogr. Mag., **36**, 1920, 183–5.

BEAUFORT, FRANCIS. 1774–1858.

Reared in an Irish country rectory, Beaufort entered the Navy and in 1825 became Hydrographer, an office he held for twenty-six years. He sailed the coasts of Australia, New Zealand, South America, the West Indies and China, and brought the technique of maritime sounding to great efficiency: his name is best known for the Beaufort Scale of wind force and the meteorological symbols still used.

J. R. Geogr. Soc., **28**, 1858, cxxiv–cxxvii.

BERGHAUS, HEINRICH. 1797–1884.

Most famed for his pioneer work in the mapping of world distributions, Berghaus inspired many workers on atlases during the nineteenth century, notably A. Petermann and the Johnston family in Edinburgh. The great *Physikalischer Atlas* is generally regarded as the foundation of all modern atlases of a regional character and its publication, along with the works of Ritter and von Humboldt, as marking the beginning of modern academic geography.

Howarth, O. J. R., in *Advanc. Sci.*, **30**, 1951, 162.

BOWMAN, ISAIAH. 1878–1950.

Bowman, like W. M. Davis, believed in publishing and in attractive presentation: he began the transformation of the *Geographical Review* into an admirably-produced journal when director of the American

Geographical Society, 1915–35, and sponsored a valuable research series. From his early travels in South America came two books and the 1:1,000,000 maps scheme; from his world studies and experience with the American Commission at Versailles his *New World*, 1919; from his knowledge of university teaching an enthusiam for graduate schools.

Geogr. J., **115**, 1950, 226–30; *Ann. Ass. Amer. Geogr.*, **40**, 1950, 335–50: *Geogr. Rev.*, **41**, 1951, 7–65. *Scot. Géogr. Mag.*, **66**, 1950, 3.

BROWN, RALPH HALL. 1898–1948.

Widely travelled in the United States, Brown found his main inspiration in its historical geography and showed a gift for acquiring the spirit of past periods. This was seen particularly in his book of 1943 published by the American Geographical Society, *Mirror for Americans: Likeness of the Eastern Seaboard, 1810*, and in the monograph, *The American Geographies of Jedidiah Morse*, published by the Association of American Geographers in 1941. His *Historical Geography of the United States* appeared in 1948.

Ann. Ass. Amer. Geogr., **38**, 1948, 305–09; *Geogr. Rev.*, **38**, 1948, 505–06.

BRUNHES, JEAN. 1869–1930.

Having studied with Vidal de la Blache, he showed great independence of mind and cared most for the study of *réalités actuelles*. Possessing great oratorical and literary powers, he was also a minutely careful researcher in his travels, and his first major work, 1902, on irrigation in Iberia and North Africa shows this clearly. His pioneer work on human geography, concerned largely with man's work on the earth, appeared in 1910, and in 1921 his *La Géographie de l'Histoire*, with C. Vallaux as joint-author, showed wide ideas. His *Géographie Humaine de la France*, 2 vols., 1920, 1926, each volume written with a collaborator, dealt with river basins rather than regions, and with topics in the later sections.

Ann. Géogr., **39**, 1930, 549–53.

BRYAN, KIRK. 1888–1950.

Born in New Mexico, he had a permanent interest in semi-arid areas, and after the 1914–18 war worked in Arizona, New Mexico and California for the Ground Water Division of the U.S. Geologic Survey. From 1926, at Harvard, his geomorphology was applied to many cognate studies, including geology, water engineering, soil science, climatology, anthropology, archaeology and human geography. Like many workers from semi-arid areas he favoured theories of environmental determinism.

Ann. Ass. Amer. Geogr., **41**, 1951, 89–94: *Geogr. Rev.*, **41**, 1951, 165–6.

BUNBURY, EDWARD HERBERT. 1811–95.

Having a training in classics, he contributed several articles to Smith, William, *Dictionary of Greek and Roman Geography*, 1870, but is best known for his great *History of Ancient Geography*, 1879. He worked on the principle of a careful study of ancient authorities, supplemented by the findings of modern travellers.

Geogr. J., **5**, 1895, 498–500.

CHAMBERLIN, THOMAS CHROWDER. 1843–1928.

From 1873, he worked as geologist for the state of Wisconsin and the four-volume survey was published by 1883: meanwhile, in 1881, he was appointed to the glacial division of the new national Geological Survey. He wrote extensively on glaciation, and the classic paper with R. D. Salisbury (q.v.) on the driftless area of Wisconsin, came out in 1885. The first formal attempt to define the multiple glaciations of the U.S.A. appeared in James Geikie's *Great Ice Age*, 1895. His later work was on broad aspects of geology, partly with R. D. Salisbury. Chamberlin went as a geologist on the Peary Antarctic Expedition of 1894.

Biogr. Mem. Nat. Acad. Sci., **15**, 1934, 307–407, by R. T. Chamberlin.

CHISHOLM, GEORGE GOUDIE. 1850–1930.

Originally a student at Edinburgh University, he spent much of his life as a university extension lecturer in London. His valuable *Handbook of Commercial Geography*, was published in 1889 and he also edited Longmans' *Gazetteer of the World*, 1895. In 1908 he became the first lecturer in Geography at Edinburgh University. Interested in a wide range of social problems, he coined the phrase 'economic ethnography' to include studies of racial, national and economic rivalries on living standards and population problems.

Geogr. J., **75**, 1930, 567: *Scot. Geogr. Mag.*, **46**, 1930, 101–04.

COLBY, THOMAS FREDERICK. 1784–1852.

The Ordnance Survey became his lifework from 1802, at first in Britain and from 1824 in Ireland, where he was responsible with Thomas A. Larcom for the six inches to one mile survey. He regarded a survey as serving a social and intellectual purpose: it could be a basis for national improvement, along with museums of economic geology. He was a keen advocate of contours and also of geological maps: he urged that in Ireland the railways should form one general system rather than disconnected lines. From 1838–46 he worked for the Scottish Survey.

J. R. Geogr. Soc., **23**, 1853, lxviii–lxx: Portlock, J. E., *Memoir of the Life of Major-General Colby*, London, 1869.

CORTAMBERT, PIERRE FRANÇOISE EUGENE. 1805-81.
Born at Toulouse, he wrote extensively on geography, partly in the *Bulletin de la Société de géographie*. His *Physiographie* of 1836 was entirely on physical geography, but his other writings show an interest in discovery and exploration, and in 1860 he produced a revised edition of the French classic, the *Géographie de Malte-Brun*. From 1860, he was in charge of the geographical section of the Bibliothèque Nationale.
Bulletin de la Société Géographique, Paris, 7ème série, 1881, 239-42.

CVIJIĆ, JOVAN. 1865-1927.
After studying in Vienna with Penck and Suess, his first book appeared in German on karsts in 1893. Appointed in the same year to the school that later became Belgrade University he began a long series of field sessions that included studies of past and present lakes, of ancient glaciers and of the people. In 1906 he published a memoir on the ethnography of Macedonia, and his fine book on the Balkans appeared in Paris in 1918. He was instrumental in fixing the boundaries of Yugoslavia.
Ann. Géogr., **36**, 1927, 181-3: *Geogr. Rev.*, **17**, 1927, 240.

DAVIS, WILLIAM MORRIS. 1850-1934.
A Harvard graduate, he worked in his twenties as meteorologist and geologist, and became internationally known as a geomorphologist by 1890 when he became professor of Physical Geology at Harvard. He spent whole sessions abroad, notably 1908-09 at Berlin. To him, geomorphology was the foundation of geography but not an end in itself. His 500 papers include many on meteorology and some on education: his conclusions are presented with a finely reasoned clarity but have been strongly criticized by his successors.
Ann. Ass. Amer. Geogr., **40**, 1950, 171-236: *Geogr. J.*, **84**, 1934, 93-5: Wright, J. K., *Geography in the Making*, 1952, 123-4.

DE GEER, STEN. 1886-1933.
Baron de Geer's first work was on geomorphology with cartography, but his main contribution was his mapping of population, first broached in an article in *Ymer*, 1908, which culminated in his atlas of Swedish population in 1919. He also wrote extensively on towns, including Stockholm, and apparently intended to consider wider aspects of human geography, such as the relation between human and physical regions. Due to his early death, his main memorial is his Atlas and the work of the later Swedish geographers to whom he gave inspiration.
Geogr. Rev., **23**, 1933, 685-6: *Svenskt Biografiskt Lexikon*.

DE LA BLACHE, PAUL VIDAL. 1845-1918.
After a university training in history and geography, he travelled in

Italy, Greece and the Near East, and in 1872 returned to Nancy as a university teacher: he moved to Paris in 1877. His book *États et Nations de l'Europe* appeared in 1889 and his atlas in 1894. With colleagues he founded the great *Annales* in 1893, and also the annual *Bibliographie*. His *Tableau de la Géographie de la France*, 1903, was a contribution to literature as well as to geography. In 1917, *France de l'Est* showed his concern for the eastern areas he knew so well, but his *Principles of Human Geography* was constructed by de Martonne from articles and an unfinished manuscript after his death.

Ann. Géogr., **27,** 1918, 161–73: *Scot. Geogr. Mag.*, **34,** 1918, 266–7.

DEMANGEON, ALBERT. 1872–1940.

Trained in history and geography, his monograph on Picardy, 1905, showed intensive fieldwork: his later works included *Le Déclin de l'Europe*, 1920 and *L'Empire Britannique*, 1923, both of which were translated into English. He produced the first two volumes in the *Géographie Universelle*, on the British Isles and Belgium, Holland and Luxemburg in 1927. An editor of the *Annales de Géographie*, he wrote a large number of papers and notes, and was working on a human geography when he died.

Ann. Géogr., **49,** 1940, 161–9: *Geogr. Rev.*, **31,** 1941, 155–6.

DE MARGERIE, EMMANUEL. 1862–1953.

Never at a university, de Margerie was from 1894 a director of the great *Annales de Géographie:* he translated Suess's *Das Antlitz der Erde* into *La Face de la Terre*, various editions from 1897–1918. His work on the Jura appeared in two parts, 1922 and 1936. He wrote articles, reviews and obituaries, and gave substantial help in the production of the 1:1,000,000 map and of geological maps of Africa and Alsace-Lorraine.

Ann. Géogr., **63,** 1954, 81–7; *Geogr. J.*, **120,** 1954, 130–1; *Geogr. Rev.*, **44,** 1954, 600–02.

DE MARTONNE, EMMANUEL. 1873–1955.

A student of Vidal de la Blache, his main interests were physical and regional geography; he served at the Universities of Rennes 1899–1905, Lyon 1905–9, and the Sorbonne, where he directed the Institut de Géographie, 1927–1944. His main works were the *Traité de Géographie Physique*, 1909 and later editions, the two volumes on *Europe Centrale* in *Géographie Universelle*, 1930–1, and the physical geography volume of France in the same series, 1942. From the death of Vidal de la Blache in 1918 to his retirement in 1945 he was the leader of the French geographical school.

Ann. Géogr., **65,** 1956, 1–14; *Geogr. Rev.*, **46,** 1956, 277–9; International Geographical Union, *Newsletter*, **7,** 1956, 3–7.

DOMINIAN, LEON. 1880–1935.

Born at Constantinople, he spent several years in Europe and Asia Minor, and lived in Malta. An expert linguist, he went to America in 1903 as a mining geologist. From 1912–17 he served on the staff of the American Geographical Society, and wrote his scholarly book, *The Frontiers of Language and Nationality in Europe* which, said W. L. G. Joerg, 'supplied the desired detailed discussion of the problems of nationalities in Europe and the Near East and their geographical setting.' He entered the consular service and died at Montevideo.

Ann. Ass. Amer. Geogr., **26,** 1936, 197–8; *Geogr. Rev.*, **25,** 1935, 687–8.

DRYER, CHARLES REDAWAY. 1850–1926.

At first a teacher of science, he worked with the Indiana Geological Survey, and from 1893–1913 was professor of geology and geography at the Indiana State Normal School. He wrote various works on general physical and economic geography, and some local studies; he did much to advance the study of geography in the United States. An admirer of A. J. Herbertson, he used regional divisions effectively in his *High School Geography*, 1911· his later papers advocated regional division on an economic basis.

Geogr. J., **70,** 1927, 509: *Geogr. Rev.*, **17,** 1927, 506.

FAWCETT, CHARLES BUNGAY. 1883–1952.

A student at Nottingham and Oxford, he worked in universities at Southampton, Leeds and London. His work on political geography showed the inspiration of Mackinder and his teaching on regional geography that of Herbertson. Of his writings, the best known is his small but suggestive *Provinces of England* 1919 and an article on conurbations (*Geogr. J.*, **79,** 1932, 100–16).

Geogr. J., **118,** 514–16: *Geography*, **37,** 1952, 232–3; *Geogr. Rev.*, **43,** 1953, 281–2.

FENNEMAN, NEVIN MELANCHTHON. 1865–1945.

Having served for many years on geological surveys, from 1907–37 he worked at the University of Cincinnati and published papers on the regional physiography of the United States, which culminated in his two volumes on this subject in 1931 and 1938. In a paper of 1936 he argued that landforms interpreted as indicating former peneplanation could be produced by normal erosive agencies without reference to a base level or a cycle. To him regional study was the core of geography.

Ann. Ass. Amer. Geogr., **35,** 1945, 181–9; *Geogr. Rev.*, **35,** 1945, 682.

FINCH, VERNON CLIFFORD. 1883–1959.

Trained partly in the graduate school at Chicago, he went to the

University of Wisconsin in 1910 and stayed till 1945. He collaborated with O. E. Baker in the writing of the *Geography of the World's Agriculture*, in 1917, and was part-author of some college texts, notably on economic geography with R. H. Whitbeck and on general aspects with G. T. Trewartha.

   Geogr. Rev., **50,** 1960, 443–5.

FITZROY, ROBERT. 1805–65.
   He organized the 'Beagle' voyages, including that of 1831–6, which led to a series of books by Charles Darwin. Eventually he became director of the Meteorological Office and established a system of sending storm signals to ports to prevent shipwrecks. As Governor of New Zealand from 1843–5, his views on the allocation of land to the Maoris were not acceptable to the government and he was recalled.
   D.N.B., *J. R. Geogr. Soc.*, **35,** 1865, cxxviii–cxxxi.

FORBES, HENRY OGG. 1851–1932.
   Educated at Aberdeen and Edinburgh universities, he eventually became a museum director at Canterbury, New Zealand, and at Liverpool. He travelled in many remote parts of the world collecting birds and fauna, and finally to Peru to advise on guano deposits. In the 1880's his difficult and largely unsuccessful attempts to enter New Guinea were closely followed.
   Geogr. J., **81,** 1933, 93–4.

FREEMAN, EDWARD AUGUSTUS. 1823–92.
   From 1845 a Fellow of Trinity College, Oxford, he became Regius Professor of History in 1884. In his work he included descriptions of historical sites, for example in his five-volume *History of the Norman Conquest*. His *Historical Geography of Europe*, with an atlas volume, 1881, was concerned partly with the boundaries of states at various periods.
   Scot. Geogr. Mag., **9,** 1893, 36–7.

FRESHFIELD, DOUGLAS WILLIAM. 1845–1934.
   A man of private means, described as 'a cultivated Victorian' he worked for the Alpine Club and the Royal Geographical Society, of which he was president from 1898–1911. Himself a mountaineer in the Alps, Caucasus and Himalayas, in 1905 he joined an unsuccessful and rain-drenched expedition to Ruwenzori. He gave help to Everest expeditions and worked for geographical education in schools and universities.
   Geogr. J., **83,** 1934, 257–62: *Geography*, **19,** 1934, 60.

GALLOIS, LUCIEN. 1857–1941.
   His first major work was *Les Géographes Allemands de la Renaissance*,

1890, but his discussion of regional aspects in France, *Régions naturelles et noms de pays*, 1908, is more widely known. He was from the 1890's a member of the editorial board of the *Annales de Géographie*, and during the 1914–18 war made studies of the Saar Basin, the frontiers of northeast France, central Greece, Salonika and Suez. He edited the *Géographie Universelle*.

Geogr. Rev., **36**, 1946, 163–4.

GEDDES, PATRICK. 1854–1932.

Having for most of his life a post at Dundee which involved teaching for only part of the year, he used his biological learning as a basis for planning, and was a perpetual source of stimulating ideas. Influenced by many French thinkers on sociology and planning, he in turn influenced many British geographers as well as workers in other fields. He left a practical memorial in several replanned cities, particularly in India and Palestine.

Essay by Lewis Mumford in Barnes, H. E. (ed.) *An Introduction to the history of Sociology*, Chicago, 1948, 677–95: Boardman, P., *Patrick Geddes, Maker of the Future*, Chapel Hill and Oxford, 1944.

GILBERT, GROVE KARL. 1843–1918.

A field geologist, Gilbert was responsible for several of the classic memoirs of the U.S. Geological Survey and notably for that on Lake Bonneville, 1890, hailed by W. M. Davis as his masterpiece. Eleven years earlier, the memoir on the Henry Mountains appeared. In his work he was strongly conscious of glacial detail and of the processes of stream erosion and he contributed much to the modern development of geomorphology, not least to the work of W. M. Davis.

Biogr. Mem. Nat. Acad. Sci., **21**, 1927, by W. M. Davis.

GREELY, ADOLPHUS WASHINGTON. 1844–1935.

In 1881–2, he was in command of the most northerly post, at 81° 44' N., on Ellesmere Island, but the American expedition was left to endure immense hardships through the failure of the relief ship to arrive. During his long subsequent career with the army, he wrote several reports on climatology, including one of 1891 on the arid regions of the U.S.A., and as Chief Signal Officer was responsible, from 1886–91, for the government's meteorological work. From 1900–04, he was on the Alaska survey for telegraph lines. His works on the Arctic were widely read.

Geogr. J., **86**, 1935, 563–4; *Geogr. Rev.*, **26**, 1936, 161.

GUYOT, ARNOLD. 1807–84.

Of Swiss birth, he became a student of theology in Swiss and German universities, and acquired an interest in natural sciences through friendship with Agassiz, the geologist: he attended the lectures of von Humboldt and Ritter and became a disciple of the latter. After some years as a private tutor in France, he taught history and geography in Neuchâtel from 1839–48 at the Academy. This was suppressed in 1848 and he followed Agassiz to America, where he wrote *The Earth and Man*, and from 1854 became professor of geography at Princeton.

Globe, **23**, Geneva, 1884, 1–72 (by Charles Faure).

HEAWOOD, EDWARD. 1864–1949.

Trained in the classics he spent almost all his working life in the library of the Royal Geographical Society. Internationally known for his writings on the history of cartography, he worked also on geographical discovery, especially in the seventeenth and eighteenth centuries. On Africa he wrote a textbook in 1896 and a chapter on its exploration, 1783–1870, in the *Cambridge History of the British Empire*, 1940.

Geogr. J., **113**, 1949, 143–4: *Geogr. Rev.*, **39**, 1949. 677–8.

HEDIN, SVEN. 1865–1952.

From the universities of Stockholm and Uppsala, he went to Berlin and Halle and acquired a rich training in meteorology, geology and zoology. He visited the Caucasus, western Persia and Mesopotamia before he was twenty-one, and remained an intrepid explorer for fifty years. In all, he published seventy-five books, of which some were translated into several languages. His claim to be the first discoverer of the sources of the Indus, Brahmaputra and Sutlej is contested.

Geogr. J., **119**, 1953, 252–3: *Geogr. Rev.*, **43**, 1953, 424–5.

HERBERTSON, ANDREW JOHN. 1865–1915.

His early work was in meteorology and oceanography and he was part-author of Bartholomew's *Atlas of Meteorology*. He studied at Freiburg (Baden), Montpellier and Paris, and in 1905 published his famous paper on natural regions. His first university appointment, in 1891–2, was as demonstrator in Botany at Dundee University College under Patrick Geddes; from 1894–6 he was at Owens College, Manchester and from 1899 at Oxford. Various textbooks were published, some with his wife.

Geogr. J., **46**, 1915, 319–20: *Geogr. Teach.*, **8**, 1915–16, 143–6: *Geography*, **21**, 1936, 18–27: *Scot. Geogr. Mag.*, **31**, 1915, 486–90.

HETTNER, ALFRED. 1859–1941.

Thanks to the advance of geography in Germany, he studied geography from his schooldays, and as a student was in contact with Wojeikoff the climatologist, Ratzel and Richthofen. He was mainly a physical and regional geographer, very widely travelled, and his various physical papers were collected in four volumes from 1919 onwards: his book on Europe originally appeared in 1907, but was revised with the addition of a volume on the rest of the world in 1923. In 1927 his much-quoted methodology book appeared, *Geographie, ihre Geschichte, ihr Wesen und ihre Methoden.*

Petermanns Mitt., **92**, 1948, 188–93.

HINKS, ARTHUR ROBERT. 1873–1945.

Known first as an astronomer of distinction, he joined the staff of the Royal Geographical Society in 1912, and during the 1914–18 war directed the production of one hundred sheets of the provisional 1:1,000,000 map of the world and some sheets of the 1:2,000,000 map of Africa. From 1919 he acted for several years as secretary to the new Permanent Committee on Geographical Names. Always in close touch with explorers, he revised the Society's *Hints to Travellers*: his writing included numerous articles and two books, *Map Projections*, 1912, *Maps and Survey*, 1913, both of which achieved later editions.

Geogr. J., **105**, 1945, 146–51.

HUGHES, WILLIAM. 1817–1876.

For many years professor of geography in King's College, London, which became part of the university, and also in Queen's College, which remained a school, he wrote various manuals on geography which ran into several editions, and also published some lectures on the general aim and purpose of geography.

J. R. Geogr. Soc., **47**, 1877, clv–clvi.

HUMBOLDT, ALEXANDER VON. 1769–1859.

Educated at Göttingen, Humboldt was for several years a mining geologist, but from 1799–1804 he travelled in Spanish America, on which his main publications appeared in 1811 and, in more extended form, from 1814–25. Apart from short visits to foreign countries, from 1804 he lived in Paris as a writer for more than twenty years; in 1828 he went to Siberia, on which his main publications was the two-volume *Asie Centrale*, published in French, 1843. His *Kosmos*, in five volumes, 1845–62, had a wider aim than his earlier works which were largely physical, including especially climate and vegetation. Towards the close of a long life, Humboldt looked to the idea of the human unity of the world.

Otté, E. C. (translator), *Cosmos*, London, 1849 (in Otté's preface): Hartshorne, R., *The Nature of Geography*, 1939, 48–88, *Perspective of the Nature of Geography*, London and Chicago, 1959: Tatham, G. in Taylor, G., *Geography in the Twentieth Century*, London, 1950, 48–59.

HUNTINGTON, ELLSWORTH. 1876–1947.

His early professional years were spent as a geologist but in 1904 he turned to climatology and its effects on life, studied with wide travels. His early works were *The Pulse of Asia* 1907 and *Palestine and its Transformation*, 1911, but from 1914 alluringly-titled books like *Civilization and Climate*, 1915, appeared: from 1924 he became concerned with heredity and eugenics, including the *Character of Races*, 1924. In the last seventeen years he wrote large textbooks, some in collaboration. He published twenty-nine books, parts of twenty-seven other books and 180 articles.

Ann. Ass. Amer. Geogr., **38**, 1948, 38–50: *Geogr. Rev.*, **38**, 1948, 153–5.

JOERG, WOLFGANG LOUIS GOTTFRIED. 1885–1952.

A son of German and Swiss parents, his education was as cosmopolitan as his background: he studied in the universities of Geneva, Gottingen and Leipzig as well as in Columbia. From 1911–37 he worked for the American Geographical Society, at first with the *Geographical Review* and later as editor of the Research volumes, which included two on Polar areas, *Pioneer Settlement*, a study of the Canadian frontiers of settlement and others. He worked on the material for the Versailles Treaty of 1919, and wrote illuminating essays on new European boundaries, atlases and geographical work in the early 1920's. In 1937 he became Chief of the Division of Maps and Charts in the National Archives, Washington.

Ann. Ass. Amer. Geogr., **43**, 1953, 255–63: *Geogr. Rev.*, **42**, 1952, 482–8: Wright, J. K., *Geography in the Making*, 1952.

JOHNSON, DOUGLAS WILSON. 1878–1944.

Beginning his academic career as a geologist, he taught at Columbia University for over thirty years, and followed in the steps of W. M. Davis but with marked originality of mind. His work on *Shore Processes and Shoreline Development* in 1919 was followed by his study in 1921 of the battlefields of the war and in 1931 by his remarkable *Stream Structure on the Atlantic Slope*. A founder and editor of the *Journal of Geomorphology*, he gave it some fine essays on the scientific method of research.

Ann. Géogr., **55**, 1946, 49–52: *Geogr. Rev.*, **34**, 1944, 317–18.

JOHNSTON, ALEXANDER KEITH. 1804–71.

As head of the map firm in his name, he was in close touch with H. Berghaus and A. H. Petermann. His atlases did much to introduce regional ideas in Britain, and in 1851 he urged greater attention to the Ordnance Survey. His son, ALEXANDER KEITH, 1844–79, was apprenticed to Stanford's, where he worked on the *Globe Atlas of Europe* and the maps for Murray's *Guide to Scotland*. In 1870 he published the *Lake Regions of Central Africa*: but he died on an expedition to Lake Nyassa.

J. R. Geogr. Soc., **42**, 1872, clxi–clxiii: *Proc. R. Geogr. Soc.*, **1**, 1879, 598–600.

KELTIE, JOHN SCOTT. 1840–1927.

Educated in Scotland, he worked for ten years with a publishing firm in Edinburgh and from 1871 with Macmillan, where he became sub-editor of *Nature* in 1873 and editor of the *Statesman's Year Book* in 1880. In 1883, he joined the staff of the Royal Geographical Society and in 1884 he embarked on a year's tour and research into geographical education which was described in the influential report of 1885. He wrote articles and textbooks but his main work was the administration of the R.G.S.

Geogr. J., **69**, 1927, 281–7: *Geogr. Rev.*, **17**, 1927, 339–40.

KÖPPEN, VLADIMIR PETER. 1846–1940.

Born in St Petersburg, he spent his youth in the Crimea and early acquired an interest in climate and vegetation. In 1875 he moved to a post at the Deutsche Seewarte in Hamburg, where he stayed till 1919. His first version of the world climatic regions was published in 1900, and was based largely on plants as indicators, but in 1918 this was modified to make it freer from botanical geography. He edited the great *Handbuch der Klimatologie*, and wrote many papers, and also the climatic section of *Die Klimate der Geologischen Vorzeit* with his son-in-law, Alfred Wegener, best known for his theory of continental drift. The Köppen climatic regions were used by A. J. Herbertson and J. G. Bartholomew in the *Atlas of Meteorology* and influenced Herbertson's work on natural regions.

Petermanns Mitt., **86**, *Geogr. Rev.*, 1940, 339: **31**, 1941, 154–5.

KROPOTKIN, PETER ALEXEIVICH. 1842–1921.

Best known in later life as a humanist, Prince Kropotkin carried out surveys in Manchuria and published a study of structural lines in Asia: he also worked on glaciation in Finland and Sweden. He escaped from Russia in 1876, lived in Switzerland, France and Britain and finally returned to Russia in 1917: his later works are on humanism and nihilism.

Howarth, O. J. R., *Advanc. Sci.*, **30**, 1951, 164.

LARCOM, THOMAS AISKEW. 1801–79.

Commissioned in the Royal Engineers in 1820, he spent almost all his working life in Ireland, where he worked with T. F. Colby (q.v.) and J. E. Portlock (q.v.) on the Ordnance Survey, which he directed from 1828–46. The Templemore Memoir (p. 45) was his conception: he arranged that the Census volumes should include occupational data and agricultural statistics from 1841. In 1846, he became commissioner for public works, and directed many of the famine relief projects. From 1853, he was Under-Secretary for Ireland. He edited the Down Survey of 1655–6.

D.N.B., J. R. Geogr. Soc., **50,** 1880, clxviii.

LOBECK, ARMIN KOHL. 1886–1958.

Much of his professional life was spent in the geology departments of Columbia and Wisconsin universities, and his main interest was the illustration of the material he presented. His work on *Block Diagrams*, 1924, and his *Geomorphology* of 1939 have made comprehensible a great deal of theory: not least, they have illumined the work of W. M. Davis and other geomorphologists.

Geogr. Rev., **48,** 1958, 584–5.

MACHATSCHEK, FRITZ. 1876–1957.

In his academic youth he knew and studied with the giants of the time—Penck, von Richthofen and Suess: he became a great geomorphologist, concerned particularly with glaciation in its widest sense. After writing on the Swiss Jura, 1905, he studied the Scandinavian mountains on which his work appeared in 1908. Later in 1912 and 1921, works on the Tian Shan and Russian Turkestan were produced. In 1938 and 1940, his *Das Relief der Erde* was published.

Petermanns Mitt., **102,** 1958, 1–5.

MACKINDER, HALFORD JOHN. 1861–1947.

Few geographers have had a more varied career, for he was principal of University College, Reading, 1892–1903, director of the London School of Economics 1904–08, and a member of Parliament from 1910–22. He laid the foundations of the Oxford school of geography by his work there from 1887–1905. His early paper of 1887 to the Royal Geographical Society, his British Association address of 1905 and his chairmanship of the Geographical Association from 1893 were valuable contributions. His *Britain and the British Seas*, 1902, was a book of the new regional approach, and later his ideas of world strategy caused widespread discussion.

Geogr. J., **110,** 1947, 94–9; vol. **113,** 1949, 47–57; *Geography*. **37,** 1951, 21–43, article by E. W. Gilbert, 'Seven Lamps of Geography.'

MARBUT, CURTIS FLETCHER. 1863–1935.

After teaching for some years, he went to Missouri University, spent three years with the Missouri Geological Survey, and was a graduate student at Harvard from 1893–5. He then returned to the Missouri University until 1910, and from 1905–10 also directed the State's Soil Survey. In 1927 he published a translation of a work by the great Russian student of soils, K. D. Glinka (1867–1927), as *Great Soil Groups of the World and Their Development*. Under Marbut's direction, about half the soils of the U.S.A. were mapped. Some of his work is in the *Atlas of American Agriculture*, and in *The Vegetation and Soils of Africa* with H. L. Shantz in 1923, published by the American Geographical Society.

Ann. Ass. Amer. Geogr., **26**, 1936, 113–23; *D.A.B.; Geogr. Rev.*, **25**, 1935, 688.

MAURY, MATTHEW FONTAINE. 1806–73.
Entering the United States Navy at nineteen, he published his *Navigation* in 1834, and in 1842 became Superintendent of the Depot of Charts and Instruments at Washington, where he produced charts and sailing directions that reduced ocean voyages in time considerably. His work on meteorology and hydrography was known to Humboldt, who regarded him as the virtual founder of oceanography: he was partly instrumental in establishing a uniform world notation for data. His book, *The Physical Geography of the Sea*, was published in 1856.

J. R. Geogr. Soc., **43**, 1873, clvii–clviii.

MILL, HUGH ROBERT. 1861–1950.
A student at Edinburgh, and a scientist by training, he met Patrick Geddes and learned to combine an interest in human affairs with a zeal for scientific investigation. He worked from 1887 at the Scottish Marine Station in Granton and also as a university extension lecturer, and in 1892 went to the Royal Geographical Society as Librarian: in 1901, he became director of the British Rainfall Organization. With Edward Heawood and A. J. Herbertson, he began a survey of English Lakes. His proposals for an official regional survey of Britain were excellent but fruitless though he had a considerable influence on private workers. He was an expert on Polar exploration, but without any practical experience of such areas.

Geogr. J., **115**, 1950, 266–7; *Geography*, **35**, 1950, 124–7; *Geogr. Rev.*, **40**, 1950, 657–60; *Nature*, **165**, 1950, 791; *Scot. Geogr. Mag.*, **66**, 1950, 1–2; *The Times*, April 6, 12, 13, 18, 1950; Mill, H. R., *An Autobiography*, London, 1951.

MURCHISON, RODERICK IMPEY. 1792–1871.
After an army career, 1808–15, he turned to physical science and for fifty years worked as a geologist; from 1855 he was Director-General of the Museums of Practical Geology. He was one of the founders of the

Royal Geographical Society in 1830 and of the British Association in 1831, and was interested in African, Arctic and Australian discovery.
J. R. Geogr. Soc., **42,** 1872, cl–clvii.

NANSEN, FRIDTJOF. 1861–1930.
He travelled, at the age of twenty-one, in northern waters to Spitzbergen, Jan Mayen and Greenland; in 1888 the *Jason* voyage showed that the Interior of Greenland was a lofty ice-sheet. But his best known journey was in the *Fram,* from 1893, which entered the ice-pack north of the New Siberian Islands and began a long drift expected to cross the North Pole: finding this impossible, he attempted to reach the Pole over the ice, but gave up at 86° 14′. The voyage showed that the Arctic was a deep basin. Later Nansen added more oceanographical work. He eventually became a diplomat and worker for refugees.
Ann. Géogr., **39,** 1930, 432–6; *Geogr. J.,* **76,** 1930, 92–5.

NEWBIGIN, MARION ISABEL. 1869–1934.
Educated as a biologist, she succeeded J. Arthur Thomson as lecturer in biology and zoology in the extra-mural medical school for women at Edinburgh. From 1902 to her death, she edited the *Scottish Geographical Magazine* and made it a leading British geographical journal. Her books included a world regional geography, a study of plant and animal geography, a work on Balkan problems written during the 1914–18 war, and a most suggestive volume, *Mediterranean Lands.*
Geogr. J., **84,** 1934, 367; *Geography,* **19,** 1934, 200; *Scot. Geogr. Mag.,* **50,** 1934, 331–3.

NORDENSKJÖLD FAMILY
Three members of this ancient Swedish family did work of special geographical interest:

(1) NILS-GUSTAV, 1792–1866. Living in Finland, he was a government inspector of mines and travelled abroad extensively; he produced a map of Finland with a memoir on the striated and polished surfaces of the country's rocks.

(2) ADOLF-ERIK, 1832–1901, his third son, studied at Helsingfors University, and became a mineralogist; he accompanied his father on a visit to the copper and iron mines of the Urals in 1853, and later, in Berlin, met various savants. In 1857, he moved to Stockholm, and in 1858 began a long series of Arctic expeditions, including several to Spitzbergen. In 1868 he reached latitude 82° 42′, and in 1875 proved that the Kara Sea was navigable, which raised hopes of a north-east passage. He wrote voluminously: after the *Vega* voyage of 1878–9, during which the boat was frozen in the ice, he wrote five volumes of scientific observations. His last

Arctic trip, to Greenland in 1883, proved that it had no ice-free heart.

(3) OTTO, 1870–1928, nephew of Adolf-Erik, carried on the family tradition of exploration, which included some fine work on glaciated areas in the southern Andes and the Magellan straits. From 1901–03 he was leader of the Swedish Antarctic Expedition, and later he visited Iceland, Spitzbergen and Greenland. His writings were mainly in German with one chapter in the American Geographical Society's *Geography of the Polar Regions*, 1928.

Proc. R. Geogr. Soc., **10**, 1865–6, 205–06; *J. R. Geogr. Soc.*, **39,** 1869, cxxxii, 131–46; Jackson, J., *Adolf-Erik Nordenskjöld*, Paris, 1880; *Geogr. J.*, **18**, 1901, 449–52; *Ann. Géogr.*, **10**, 1901, 464; *Geogr. Rev.*, **18**, 1928, 689; *Scot. Geogr. Mag.*, **17**, 1901, 595–7.

OGILVIE, ALAN GRANT. 1887–1954.

Born in Edinburgh but mainly educated at Westminster and Oxford, Ogilvie went to Berlin and Paris as a post-graduate student. After two years teaching in Oxford with A. J. Herbertson, his war service of 1914–18 gave him an abiding interest in the Balkans. With later travel and work, this led to his posthumous book, *Europe and its Borderlands*, 1957. He first visited U.S.A. in 1912 with the transcontinental geographical excursion and spent 1920–3 with the American Geographical Society, where he wrote the *Geography of the Central Andes*, 1922. He edited and partly wrote *Great Britain: Essays in Regional Geography*, 1928.

Geogr. J., **120**, 1954, 258–9; *Geogr. Rev.*, **44**, 1954, 442–4; *Scot. Geogr. Mag.*, **70**, 1954, 1–5; Miller, R. and Watson, J. W. (eds.), *Geographical Essays in Memory of Alan Grant Ogilvie*, London, 1959, xi–xvi, 1–6.

PASSARGE, SIEGFRIED. 1867–1958.

Born in Königsberg, he travelled in the Cameroons in 1893–4, in the Kalahari and adjacent areas from 1896–9, and in Venezuela, where he led an expedition to the middle Orinoco valley, from 1901–02: in 1906–07 he was in Algeria and the Sahara. He held academic posts in Berlin and Breslau, and at the Colonial Institute, Hamburg. To some extent influenced by Richthofen, he was an original thinker: his work on geomorphology led him to general regional geography. Often in conflict with W. M. Davis, he wrote several books including *Die Kalahari*, Berlin, 1904; *Physiologische Morphologie*, Hamburg, 1912; *Vergleichende Landschaftskunde*, 4 vols., Berlin 1921–30, and *Die Erde und ihr Wirtschaftsheben*, Hamburg 1927.

Deutsches Kolonial Lexikon: Encyclopaedia Italiana XXVI (by R. Almagia).

PEARY, ROBERT EDWIN. 1856–1920.

Attracted to Greenland by a casual paper, Peary's first expedition, in 1886, was of a reconnaissance nature. After a period of naval service, he went to north Greenland in 1891 and established its insularity. Three further expeditions in the next ten years had varying success, though in 1905–06 he reached 87° 6′ but had to turn back: his final and successful expedition was in 1908–09 but following his return a fierce contest developed with another claimant to be the first discoverer of the North Pole. Peary contributed to hydrography, meteorology and the technique of exploration.

Geogr. J., **55**, 1920, 405–08; *Geogr. Rev.*, **9**, 1920, 161–9; *D.A.B.*

PETERMANN, AUGUST HEINRICH. 1822–78.

A pupil at the Geographical Art School in Potsdam, founded by Berghaus, he worked as a cartographer for A. von Humboldt. In 1845, he went to the Johnston firm in Edinburgh and later to London where he compiled some remarkable population maps, including those in the 1851 Census. He returned to Gotha in 1854, and was responsible for the Stieler *Atlas* and for the *Mitteilungen* which still bears his name.

Proc. R. Geogr. Soc., **1**, 1879, 133–4.

PORTLOCK, JOSEPH ELLISON. 1794–1864.

His early military career was in Canada, but in 1824 he joined the Ordnance Survey, then under Col. T. F. Colby, and with others he organized the triangulation of Ireland, the measurement of sea-level by tidal observations, and the general levelling of the country. He was partly responsible for the 1837 memoir on the Londonderry area. After some years as a surveyor, he returned to military duties, and held commands in Corfu, and later in Portsmouth and Cork. He wrote a memoir of his distinguished brother-officer, T. F. Colby.

D.N.B; J. R. Geogr. Soc., **36**, 1864, cxv–cxvii.

POWELL, JOHN WESLEY. 1834–1902.

Through family circumstances, he had a somewhat broken education and migratory childhood, and his career was delayed by service in the Civil War. From 1867 he began the exploration of Colorado and the Uinta mountains, on which he published reports in 1875 and 1876. He stressed the idea of antecedent rivers, and laid many of the foundations on American geomorphology. The 1879 *Report on the Arid Lands*, by various authors, was a classic and sensible assessment of the possibilities. From 1880 he was director of the Geological Survey. He viewed human evolution as cultural advance, not as 'the survival of the fittest'.

Biogr. Mem. Nat. Acad. Sci., **8**, 1919, 11–83 (by W. M. Davis).

RATZEL, FRIEDRICH. 1844–1904.

A student at several German universities, he travelled widely in

Europe, the United States, Mexico and Cuba, and worked in the universities of Munich and Leipzig. In 1878 he published a book on North America, but his most famous books are *Anthropogeographie* (1882 and 1899) and *Politische Geographie*, 1887. There are other books of a more popular character. His name has been associated, excessively, with determinist views.

Ann. Géogr., **13**, 1904, 466–7; *Geogr. J.*; **24**, 1904, 485–7; *Scot. Geogr. Mag.*, **20**, 1904, 597.

RECLUS, JEAN JACQUES ÉLISEE. 1830–1905.
Reclus was a student of Karl Ritter at Berlin in 1849. Partly through temporary political banishment, he travelled widely and began to issue his *Géographie Universelle; la Terre et les Hommes* in weekly parts from 1876. The last volume of nineteen, each with 800–900 pages, appeared in 1894, and several were revised and reissued. One aspiration of this enterprising man was to make a world globe on the 1 to 1,000,000 scale.

Ann. Géogr., **14**, 1905, 373–4; *Geogr. J.*, **26**, 1905, 337–43.

RICHTHOFEN, FERDINAND FREIBEUR VON. 1833–1905.
From his first researches in the Tyrol, he extended his range to Transylvania, and in 1859 went to eastern Asia, where he spent many years: from 1868–72 he was in China and from 1877 his material was published with the patronage of the Emperor William. In 1882, his volume on North China, with an atlas, appeared. He showed the existence of the Shantung coalfield, and the strategic possibilities of Kiaochow. He served in the universities of Bonn, Leipzig and Berlin.

Geogr. J., **26**, 1905, 679–82.

RITTER, CARL. 1779–1859.
After two years at Halle University, in 1798 he became tutor to the Hollweg family of Frankfurt and toured Europe, on which he wrote a textbook, 1804 with a small atlas volume, 1806. His *Erdkunde* was first published from 1817, but an enlarged edition, of some 20,000 pages on Africa and Asia—the latter unfinished—appeared from 1832. Europe was not included. His views of geography have been widely discussed: among his pupils was A. H. Guyot.

Scot. Geogr. Mag., **75**, 1959, 153–63 (by K. A. Sinnhuber), *J. Amer. Geogr. and Stat. Soc.*, **2**, 1860, 25–63 (by A. H. Guyot).

ROMER, EUGENIUSZ. 1871–1954.
Born in Lwow, he studied in Cracow University and later with Albrecht Penck at Vienna; he was widely travelled and studied glaciers in the Alps and in Alaska, though his main fieldwork was on the eastern Carpathians and the Tatra. But he was most famed for his

Atlas of Poland, 1916, for his work at the Peace Conferences, and for later cartographical work. He was professor of geography at Lwow, 1911–31, where in 1921 he established a cartographical institute, now moved to Cracow.

Ann. Géogr., **63**, 1954, 473; *Geogr. Rev.*, **44**, 1954, 602–03.

ROXBY, PERCY MAUDE. 1880–1947.

Trained as an historian in Oxford, he went to Liverpool University in 1904 and stayed until his retirement forty years later, when he settled in China as British Council representative. His main work was on China, but he also wrote critically on regionalism and on the aims and methods of human geography. In all his thinking, he combined historical with geographical strands. He was an orator of commanding presence, and did much to develop geography teaching in Britain and abroad.

Geogr. J., **109**, 1947, 155–6; *Geogr. Rev.*, **37**, 1947, 506; *Nature*, **159**, 1947, 462.

SALISBURY, ROLLIN D. 1858–1922.

Note—'D' was only an initial added for effect!

Best known for a college text on 'Physiography', 1907, with shorter editions later for school use, Salisbury had worked with T. C. Chamberlin on the driftless area of Wisconsin and neighbourhood. He organized the first geography department of some size in U.S.A. at Chicago, but after working there from 1903–19 he moved to the geology department. With Chamberlin he published a three-volume *Geology*, 1904–06: he was not attracted by the 'cause and effect' ideas of W. M. Davis, and distinguished firmly between physical and human geography.

Ann. Ass. Amer. Geogr., **43**, 1953, 4–11; **47**, 1957, 276; *Geogr. Rev.*, **12**, 1922, 659.

SCHRADER, FRANZ. 1844–1924.

Having no university training, Schrader became a fine cartographer and compiler of atlases. He spent the leisure of sixty years surveying the southern slopes of the Pyrenees, on which he produced a classic map, 1 : 800,000. From 1891–1913 he edited the *Année Cartographique*, which dealt with explorations and boundary changes over the whole world. With various collaborators, he produced the *Atlas Moderne*, 1889; the *Atlas de Géographie Historique*, 1896; the *Atlas Universelle*, first edition 1881–1911. Written out of long experience, his 'Foundations of Geography in the Twentieth Century', a Herbertson Memorial lecture, reveals the man and the time.

Ann. Géogr., **34**, 1925, 564–7; *Geogr. Teach.*, **10**, 1919–20, 44–53 (the lecture mentioned above).

SCOTT, ROBERT FALCON. 1868–1912.
A naval officer and polar explorer, his first expedition was in the *Discovery*, 1901–04, which reached the Ross Sea. In 1910 he left in the '*Terra Nova*' and with four companions reached the South Pole in January 1912, one month after R. Amundsen. The Scott Polar Institute in Cambridge was founded as a memorial and a permanent centre for Polar research.
D.N.B., *Geogr. J.*, **41**, 1913, 201–22.

SEMPLE, ELLEN CHURCHILL. 1863–1932.
Born into a cultivated family, Miss Semple as a student read widely in sociology and economics as well as in geography. In 1891–2 she went to Leipzig as a student of Ratzel, and from 1897 she devoted most of her time to writing, though she taught intermittently at several American universities. Her first book, *American History in its Geographic Conditions*, 1903, is still recognized as significant but she is even better known for the *Influences of Geographic Environment*, 1911, a modification of the theories of Ratzel. Her later work, from 1915, was on the Mediterranean.
Ann. Ass. Amer. Geogr., **33**, 1933, 229–40; *Geogr. Rev.*, **22**, 1932, 500–01.

SHACKLETON, ERNEST HENRY. 1874–1922.
Of Anglo-Irish stock, he entered the Navy, and sailed with R. F. Scott on the *Discovery* in 1901 and reached 82° 16′ 33″ S. In 1907, in the *Nimrod*, he reached the Ross Barrier and, in 1909, the South Magnetic Pole. The *Endurance* voyage of 1914–16 was rich in results; for nine months the ship was frozen into the pack-ice and drifted. In September 1921, Shackleton sailed on the *Quest* but he died suddenly in January 1922.
D.N.B.; *Geogr. J.*, **59**, 1922, 228–30 and **61**, 1923, 133–5; *Geogr. Rev.*, **12**, 1922, 313 and **13**, 1923, 158–60.

SHANTZ, HOMER LEROY. 1876–1958.
Born in Colorado, Shantz became a botanist with a special interest in the vegetation of semi-arid lands. After the 1914–18 war his interests turned to Africa, on which he wrote a series of articles in *Economic Geography*, 1940–43. His paper in *Ann. Ass. Amer. Geogr.*, **13**, 1923, on the natural vegetation of the Great Plains is classic, and so too is the work on natural vegetation in the *Atlas of American Agriculture*. Shantz in his last years was concerned with the great changes he had seen in Colorado, the Great Plains and Africa.
Geogr. Rev., **49**, 1959, 278–80.

SÖLCH, JOHANN. 1883–1951.
A student at Vienna, he later went to study with Penck and von

Richthofen in Berlin, and also with J. Partsch in Leipzig. He spent his teaching years in the universities of Graz, Innsbruck and Vienna and much of his work is on the geomorphology of the Alps, though he worked on to the human applications. His last work, of which the second volume appeared posthumously, was *Die Landschaften der Britischer Inseln*, 1951 and 1952.

Petermanns Mitt., **96,** 1952, 110–14.

SOMERVILLE, MARY. 1780–1872.

Acquainted during her long life with many of the great writers in science and the arts of her time, Mrs Somerville published in 1836 a work *On the Connexion of the Physical Sciences*, which included a discussion of tides, currents, climate, plant geography and other natural phenomena. Her *Physical Geography* was first published in 1848 and went into several editions. In 1879 the college which bears her name in Oxford was founded as a Hall for women.

Baker, J. N. L., in *Geogr. J.*, **III**, 1948, 207–22.

SPEKE, JOHN HANNING. 1827–64.

Serving with the Indian Army, he used his leave periods to explore the Himalayas and Tibet. From 1854 he joined several African expeditions and his name is inseparably associated with discoveries on the Upper Nile, with Lakes Tanganyika and Victoria Nyanza.

J. R. Geogr. Soc., **35,** 1865, cix–cxi.

STANLEY, HENRY MORTON. 1840–1904.

Born James Rowlands at Denbigh, he took the name of an American benefactor and first went to Africa in 1867 for the *New York Herald*. In 1871 he found Livingstone and returned to England with Livingstone's journals. Stanley's main exploration was done in 1874–7 in the upper Congo and the lake area (Victoria Nyanza, Albert Edward, Tanganyika). From 1879–84 he was in the Congo, which in 1885 became a state under King Leopold of the Belgians. He lectured in Britain, America and Australia and from 1895–1900 sat in the British Parliament.

Geogr. J., **24,** 1904, 103–06.

SUESS, EDUARD. 1831–1914.

Born in London, where his father was an evangelical pastor, he studied in the universities of Prague and Vienna, where he served for many years as professor of geology. His most widely known work is the four-volume *Das Antlitz der Erde*, 1883–1909, translated into English and French. His book on the Alps was published in 1875. Strongly socially-minded, he served on the city council of Vienna and in parliament.

Ann. Géogr., **23-4,** 1915, 371–3; *Petermanns Mitt.*, **60,** 1914, 339.

TELEKI, GRAF PAUL. 1879–1941.

Originally trained in political science at Budapest, he became an assistant in geography, and in 1909 published an atlas showing the history of cartography in the Japanese islands. He gave practical service as a geographer in the 1914–18 war, and became an expert on ethnographic distributions and minority problems. His final large work was a geography of Hungary in three volumes, 1936–9. At the time of his death he was Premier of Hungary.

Geogr. Rev., **31**, 1941, 514–15; Petermanns Mitt., **87**, 1941, 291–4.

WARD, ROBERT DE COURCY. 1867–1931.

Partly influenced by W. M. Davis, he devoted his life to the study of climatology and worked in the Department of Geology and Geography at Harvard. His early studies were on sea breezes and local thunderstorms, and his later works included *Climate, Considered Especially in Relation to Man*, 1908, *The Climates of the United States*, 1925 and the volume on the U.S.A. in the five-volume *Handbuch der Klimatologie*, by W. Köppen and R. Geiger. He saw climate in terms of the entire weather characteristics, with strong human applications. A worker for the restriction of immigration, his views contributed to the quota legislation of the U.S.A.

Ann. Ass. Amer. Geogr., **22**, 1932, 29–43.

WHITTLESEY, DERWENT STANITHORPE. 1890–1956.

Drawn to geography from the study of history, Whittlesey settled at Chicago after the 1914–18 war among a fine group of scholars, and moved to Harvard in 1928. He collaborated with W. D. Jones on an economic geography text in 1925, and later became deeply interested in political geography, as shown in his *Earth and the State*, 1939. His view was that regional work must retain a world, or at least continental view, yet be concerned with the microgeographical aspects also: he was a traveller in Africa and a detailed field-worker in Boston and district.

Geogr. Rev., **47**, 1957, 443–5.

YOUNGHUSBAND, FRANCIS EDWARD. 1863–1942.

Posted to an Indian regiment, he was sent on a reconnaissance of the Indus and the Afghan border, and then attached to the Intelligence Department to revise the gazetteer of Kashmir. In 1886 he travelled from Peking through Manchuria and crossed central Asia back to India, and his explorations of the following five years included the Karakoram and Pamir mountains. In 1904 he entered Lhasa to negotiate a treaty with Tibet, and in 1910 his work *India and Tibet* appeared. He also served as Resident in Kashmir. During his retirement he helped to promote Everest exploration.

Geogr. J., **100**, 1942, 131–7; *Geogr. Rev.*, **32**, 1942, 681.

INDEX

Aberdare, Lord (fl. 1882), 56–58
Administrative areas, 69, 119–21, 124, 126, 138, 197, 202, 228, 259
Admiralty, British, 26, 27, 29
Africa:
  commerce and colonization, 46, 57–59
  exploration, 17, 29, 53–54, 56, 58–59
  International Commission of 1876, 52
  regions of, 73
Agassiz, L., 40
Agriculture, 23, 151, 153–5, 159–60, 164–72, see also *Land*
Ahlmann, H. W., 20, 114
Air photography, 169, 252
Air routes, 14, 71, 157
Alaska, 114, 115, 157
America, United States of:
  agricultural regions of, 129–33, 140, 153, 166, 168
  Atlases of, 239–41
  city growth in, 148, 154, 179, 197–203, 251, 264
  exploration, 47, 53, 54, 247
  Geological Survey, 42, 85–86, 97, 103, 115
  industry of, 162–3
  regions of, 122–6, 128, 129
American geography:
  cartography, 103–04, 239–40
  economic, 152–3, 157, 165–8
  and environmental determinism, 74–82
  and geomorphology, 41, 66–67, 71–73, 97–109, 117, 248
  historical, 239–40, 252
  modern specialization, 22, 182–3, 185, 195, 260
  political, 205–06, 210–11, 213–14, 220–22
  regional, 122–33, 137
  societies, 49, 51, 53, 66–67, 100, 114–15, 203, 205–06, 221, 256
  urban, 193–9, 201
Ancel, J., 208

Annales de géographie, 17, 62, 63, 255–6
Antarctic:
  exploration of, 29, 43, 67, 68, 70–71, 114
  glacial recession in, 113
Anthropology, 39, 87, 177, 180–2, 206
Archaeology, 39, 177, 182
Arctic:
  exploration, 29, 43, 50, 53, 56, 67, 233
  seas, 53, 113, 233
Arrowsmith, J., 43, 55, 232, 239, 303
Association of American Geographers, 86, 246
Atlases:
  General, 32, 34, 40, 43–45, 55, 60–61, 229–30, 233–40, 256–7
  National, 240–5
Atwood, W. W., 99, 303
Australasia, Geographical Society of, 57
Australia:
  agriculture, 130, 133
  exploration, 43, 57
  mapping of, 43, 241
  settlement, 19, 26–28, 64–65, 71, 254, 262
Austria, 21, 216, 222
Austria-Hungary, 213–14, 216, 219, 222

Baker, J. N. L., 28, 38
Baker, O. E., 127–8, 165–6, 303–04
Bartholomew, J. G., 45, 82, 304
Bartholomew, map firm, 60, 67, 238
Bates, H. W., 38
Baulig, H., 104–05
Beaufort, F., 29, 304
Beaver, S. H., 263
Belgium, 128, 200, 215, 241, 243, 264
Benyon, E. D., 198
Beresford, M., 252
Berghaus, A. H., 34, 44, 45, 60, 229, 235–7, 304
Berlin Geographical Society, 27, 50

327